Handbook for
Matrix Computations

Frontiers in Applied Mathematics

Managing Editors
for Practical Computing series

W.M. Coughran, Jr.
AT&T Bell Laboratories
Murray Hill, New Jersey

Eric Grosse
AT&T Bell Laboratories
Murray Hill, New Jersey

Frontiers in Applied Mathematics is a series of monographs that present new mathematical or computational approaches to significant scientific problems. Beginning with Volume 4, this series will reflect a change in both philosophy and format. Each volume will focus on a broad application of general interest to applied mathematicians as well as engineers and other scientists.

This unique series will advance the development of applied mathematics through the rapid publication of short, inexpensive monographs that lie on the cutting edge of research. The first volume in the new softcover format is in the Practical Computing series. Future volumes will include

fluid dynamics
solid mechanics
nonlinear dynamical systems

materials science
control theory
mathematical physics.

Potential authors who are now writing or plan to write books within the scope of the series are encouraged to contact SIAM for additional information.

Publisher
SIAM
3600 University City Science Center
Philadelphia, PA 19104-2688
(215) 382-9800

Handbook for
Matrix Computations

Thomas F. Coleman
Charles Van Loan
Cornell University

Philadelphia 1988

Library of Congress Catalog Card Number 88-61637.
ISBN 0-89871-227-0

Second printing July 1989.
Third printing April 1991.

Contents

v

Preface

This handbook can be used as a reference by those actively engaged in scientific computation. It can also serve as a practical companion text in a numerical methods course that involves a significant amount of linear algebraic computation. The book has four chapters, each being fairly independent of the others.

Our treatment of Fortran 77 in Chapter 1 involves a much stronger emphasis on arrays than is accorded by other authors. We also assume that the reader has experience with some high-level programming language. This might be in the form of a recent course in Pascal or a course in Fortran taken many years ago and now half-forgotten.

The second chapter is about the Basic Linear Algebra Subprograms (BLAS). The elementary linear algebra that underpins the BLAS makes them a good vehicle for acquainting the beginning student with modular programming and the importance of "thinking vector" when organizing a matrix computation.

Chapter 3 is concerned with LINPACK, a highly acclaimed package that is suitable for many linear equation and least square calculations. The last chapter is about MATLAB, an interactive system in which it is possible to couch sophisticated matrix computations at a very high level.

A one-semester course in matrix algebra (or the equivalent) is required to understand most of the text.

Because the book spans several levels of practical matrix computations, it can fit into a number of canonically structured numerical methods courses. At Cornell we use Chapters 1 and 2 in our one-semester introductory numerical methods course. In this course it is assumed that the students are acquainted with Pascal. That is why our treatment of Fortran is brisker than what would be found in a "pure" Fortran text. In our graduate-level numerical analysis courses we use Chapters 2, 3, and 4 heavily, with Chapter 1 serving as a reference.

The BLAS and LINPACK are in the public domain and are distributed at cost through Argonne National Laboratory. MATLAB is available from MGA Inc., 73 Junction Square Dr., Concord, MA 01742.

We are indebted to Nick Higham, Bill Coughran, and Eric Grosse for catching numerous typographical errors and for making many valuable suggestions.

Thomas F. Coleman
Charles Van Loan

Chapter 1

A Subset of Fortran 77

Our treatment of Fortran 77 (F77) assumes that the reader is already familiar with some high-level language, e.g., PASCAL. There are many books devoted to the presentation of the basics of Fortran 77(see, e.g., Zwass [8]). We do not attempt to be exhaustive at this level. For example, we say nothing about the "opening" and "closing" of files, and character manipulation is mentioned only briefly. Rather, our emphasis is on matrix computations and we have attempted to be complete in this regard. We pay special attention to arrays since the implementation of matrix algorithms is an underlying theme of the book.

We believe that it is still important for students with a serious interest in scientific computation to be familiar with Fortran. This is not to say that we are advocating that all scientific computing be done in Fortran. On the contrary, the most appropriate language for a particular application at hand should be used. The language "C" is an increasingly popular choice. MATLAB is well suited to dense matrix computations and graphical work. However, we strongly believe that high-quality subroutines should be used as building blocks whenever possible. For reasons of portability and standardization, this software is almost always Fortran software and will probably continue to be so for many years. Therefore, familiarity with Fortran is essential.

A scientific programmer working in a language other than Fortran can often use Fortran subroutines directly, provided the language and computer system support such an interface. Otherwise a direct translation can be used, provided that extreme care is taken. In either case, Fortran knowledge is required.

We make two final observations about the use of Fortran. First, Fortran should seriously be considered when developing general software for a basic mathematical computation with widespread applicability. Fortran programs that adhere to professionally set "standards" are easily ported to other computing systems. This is not true for other languages. Second, it is important to realize that Fortran is not a "dead" programming language. The "modernization" of Fortran (and the subsequent updating of the "standards") is an ongoing process. Fortran 77 is now widely used and accepted (replacing Fortran 66). Fortran 8X is currently being developed. Each new Fortran basically inherits the old version as a subset to which the extensions add power and flexibility.

1.1 BASICS

We begin the discussion with a Fortran program that computes the surface area of a sphere from the formula $A = 4\pi r^2$:

```
      program area
      real r, area
c
c     This program reads a real number r and
c     prints the surface area of a sphere that
c     has radius r.
c
      read(*,*)r
      area = 4.*3.14159*r*r
      write(*,*)area
      stop
      end
```

The purpose of each line in this program is quite evident. Those lines that begin with a "c" are *comments*. The **read** and **write** statements perform input and output. The computation of the surface area takes place in the arithmetic assignment statement "area = ...". The beginning and end of the program are indicated by the **program** and the **end** statements. Execution is terminated by the **stop**. The memory locations for the variables used by the program are set aside by the line "real r, area".

In this section we elaborate on these and a few other elementary constructs.

Program Organization

A Fortran program generally consists of a main program (or "driver") and several subprograms (or "procedures"). Typically the main program and a few of the subprograms are written by the user. Other subprograms may come from a "library." Subprograms are discussed later.

A main program is structured as follows:

3

program { *name* }

{ *declarations* }

{ *other statements* }

stop
end

Each line in a Fortran code must conform to certain column position rules.

Column 1	:	Blank unless the line is a comment.
Columns 2-5	:	Statement label (optional).
Column 6	:	Indicates continuation of previous line (optional).
Columns 7-72	:	The Fortran statement.
Columns 73-80	:	Sequence number (optional).

Comments

A line that begins with a "c" in the first column is a comment. Comments may appear anywhere in a program and are crucial to program readability. Comments should be informative, well written, and sufficiently "set off" from the body of the program. To accomplish the latter begin and end each comment block with a blank comment as in the surface area program above.

Sequence Numbers

Every line in a program can be numbered in columns 73-80. This used to be a common practice but in the age of screen editors sequence numbering has become less useful for debugging.

List-directed "read" and "write"

Until we cover input/output (I/O) in depth in §1.7, we rely upon two elementary I/O constructs:

read(*,*)   { *list-of-variables* }

write(*,*)   { *list-of-variables* }

Thus,

```
read(*,*) x, y, z
```

reads three items of data and stores them in the variables named x, y, and z. Similarly,

```
write(*,*)a, b
```

prints the contents of the variables named a and b.

Messages (enclosed in quotes) may be interspersed with variable names in a **write** statement. Thus, if a and b contain 2 and 3, respectively, then

```
write(*,*)'a = ', a,'b = ', b
```

produces the output

```
a = 2.0000    b = 3.0000
```

An important detail suppressed in our I/O discussion is where a **read** physically obtains its data and where a **write** physically sends its output. The "asterisk" notation in a **read** or a **write** specifies a default device, e.g., the keyboard, the terminal screen, etc. How the default devices are set depends upon the system and it is necessary to obtain local instructions for their use.

Names in Fortran

Names in Fortran must involve no more than six characters chosen from the alphanumeric set

```
ABCDEFGHIJKLMNOPQRSTUVWXYZ0123456789
```

The name must begin with a letter. In our examples we do *not* distinguish between upper and lower case. However, it should be noted that a few F77 compilers accept only uppercase input.

Types and Declarations

Every variable in a Fortran program should be defined in a *declaration*. This establishes the variable's *type*. There are several possibilities:

integer   { *list-of-variables* }

real   { *list-of-variables* }

double precision   { *list-of-variables* }

complex   { *list-of-variables* }

logical   { *list-of-variables* }

character   { *list-of-variables* }

Thus, a program may begin with the following declarations:

```
double precision  a, b, c
integer n
complex z1, z2
double precision length, width
logical test
character*20 name
```

There are a few rules to follow regarding the naming and typing of variables.

- Variable names must begin with a letter and must be no more than six alphanumeric characters in length.

- Each variable should be declared exactly once. Automatic typing occurs otherwise. This means that variables whose names begin with letters i through n are integers and all others are real.

- It is legal to have more than one declaration per type in a program.

- Declarations must be placed at the beginning of a program, before any executable statement. This is because their purpose is to set aside memory locations for the variables used by the program.

We discuss the integer, real, double precision, and character types now. Logical and complex variables are discussed in §1.2 and §1.8, respectively.

Integer Variables

Numbers stored in integer variables are represented in *fixed point* style. In a typical computer 32 bits, $b_0, b_1,..., b_{31}$, might be allocated for each integer variable x with the convention that x has the value

$$x = (-1)^{b_{31}} \times (b_{30}b_{29}...b_1b_0)_2 \quad b_i \in \{ 0,1 \} .$$

The notation $(\cdot)_2$ is amply illustrated by the example

$$(01101)_2 = 1 \times 2^0 + 0 \times 2^1 + 1 \times 2^2 + 1 \times 2^3 + 0 \times 2^4 = 13$$

Note that because of the finiteness of an integer "word" there is an upper bound on the size of the integers that can be represented. In the 32-bit example, only integers in the interval $[-m , m]$, where $m = 2^{31} - 1 \cong 2 \times 10^9$, can be represented.

Floating Point Variables

Numbers that are stored in real or double precision variables are represented in *floating point* style. The floating point word is partitioned into a mantissa part, m, and an exponent part, e, with the convention that the value of the variable is specified by $m \, 2^e$. The length of a floating point word and how it is partitioned into the exponent and mantissa parts depends upon the computer used. A typical 32-bit floating point number x might have a 24-bit mantissa m and an 8-bit exponent e, with the convention that

$$m = (-1)^{b_{23}} \times (.b_0 b_1 \cdots b_{22})_2$$

and

$$e = (-1)^{b_{31}} \times (b_{30} \cdots b_{24})_2$$

with the convention that $x = m2^e$. Stipulating that $b_0 \neq 0$ makes the representation of a given floating point number unique. (An exception to this rule is required when $x = 0$.)

Double precision variables represent numbers in the same fashion as do real variables, but more space is allocated per variable. For example, if 32-bit words are used for real variables then typically 64-bit words would be used for double precision variables. This leaves more bits for mantissa specification, e.g., 56 bits instead of 24.

Regardless of precision, there are finitely many floating point numbers, and *rounding errors* generally arise with every arithmetic operation. Moreover, an arithmetic operation (such as a divide by a very small number) may lead to a number that is "too big" to represent in the floating point system. *Overflow* results, a situation that usually leads to program termination. Some of the hazards of floating point computation are discussed in §1.9.

Arithmetic Assignment

An arithmetic assignment statement has the form

{ *variable name* } = { *expression* }

The expression on the right-hand side is evaluated according to precise rules and the result is stored in the memory location corresponding to the variable on the left-hand side. The values of the variables on the right-hand side do not change as a consequence of the assignment. For example,

```
A = pi*r**2
```

would compute the area of the circle and store the result in the variable A, assuming that pi and r are appropriately initialized. An asterisk denotes multiplication, whereas a double asterisk specifies exponentiation.

A more complicated assignment statement is

```
root = -b + sqrt(b*b - 4.*a*c)/2.*a
```

The expression to the right of the "=" involves one of Fortran's numerous "built-in" (or intrinsic) functions, the square root. We introduce various built-in functions throughout the text. A complete list is given in Appendix 1.

Readers familiar with solving quadratic equations will recognize that the above assignment statement does *not* compute a zero of $ax^2 + bx + c = 0$. The order in which the operations are to be performed is not correctly specified. Indeed, the above assignment statement is equivalent to

```
root =  -b + ( (sqrt(b*b - 4.*a*c)/2. )*a )
```

The problem is one of *precedence*. Unless overridden by parentheses, an exponentiation (**) has a higher precedence than a multiplicative operation such as "/" or "*" which in turn has a higher precedence than an additive operation such as "+" or "-". Thus

```
w = x + y/z**2      ⟺      w =   x + ( y/(z**2) )
```

If a choice has to be made between two multiplicative operations or two additive operations, then the leftmost operation in the expression is performed first. On the other hand, repeated exponentiations are processed from right to left. These examples should clarify the possibilities:

```
w = x + y + z       ⟺      w = (x + y ) + z
w = x/y*z           ⟺      w = (x/y)*z
w = x**y**z         ⟺      w = x**(y**z)
```

Parentheses should be used to prescribe the correct order of computation to both the compiler and the reader in ambiguous cases. Thus,

```
root = ( -b + sqrt(b*b - 4.*a*c) )/(2.*a)
```

correctly assigns a zero of $ax^2 + bx + c = 0$ to the variable `root`, provided the argument passed to the square root function is nonnegative at the time of execution.

The readability of an arithmetic assignment statement is enhanced by inserting blanks in appropriate places. Thus, the above assignment is more readable than

```
root = (-b+sqrt(b * b-4. * a * c))/(2.*a)
```

In general, it is good to insert blanks around the "=", "+", and "-" signs. An exception is the unary minus sign, as

```
a = -a
```

is better than

```
a = - a
```

Complicated Expressions

Fortran statements must appear between columns 7 and 72 (inclusive) of the input record. A long arithmetic assignment may not "fit" into this space. If this is the case then it can be continued on the next line, provided that any nonzero character is placed in column six. We use the ampersand "&" for this purpose. Thus, the single line assignment

```
a = b - c - d - e - f - g - h - k
```

is equivalent to

```
a = b - c - d - e - f - g -
&                h - k
```

assuming that the "&" is situated in column 6. When spreading a statement over two or more lines, break the statement in a natural place, i.e., not in the middle of a variable name.

String Manipulation

The manipulation of string data is not particularly central to traditional scientific computation. However, it is worth knowing how F77 handles strings. To begin with, a string is a sequence of characters, e.g., 'Fortan 77!' . Variables that hold strings must be declared to be type **character**. Here is a program that sets up a string variable, assigns it a value, and then prints the result:

```
program string
character *20 str
str = 'Fortran 77!'
write(*,*)str
stop
end
```

A character variable has a size that is determined in the declaration. The "20" in "character*20" means that str can hold strings comprised of 20 or fewer characters. If no number is specified, the default length is 1, e.g.,

```
character str
```

Literal strings are surrounded by quotes when they appear in an expression. The string computations that are supported by F77 include concatenation, substring assignment, and substring extraction. For example, if s1 = 'abcde' and s2 = 'xyz' then

concatenation: s = s1 // s2   ⟹   s = 'abcdexyz'
substring assignment: s(2:4) = '123'   ⟹   s = 'a123e'
substring extraction: s = s1(2:5)   ⟹   s = 'bcde'

The length of a string and the detection of specified substrings is handled with the built-in functions **length** and **index**. Simple examples suffice to define how these functions work. If s = 'abcdabcd' then

```
k = length(s)        ⟹     k = 8
k = index(s,'da')    ⟹     k = 4
k = index(s,'cd')    ⟹     k = 2
k = index(s,'dc')    ⟹     k = 0
```

The **length** function returns the length of the string named, not the length of the string variable that contains the string. If s = " , the empty string, then **length** returns zero. The **index** function also returns an integer. In a reference of the form

$$k = \text{index}(s1, s2)$$

index looks for an occurrence of the second string in the first string. If no occurrence is found, then zero is returned. Otherwise, the "starting" slot of the first occurrence is returned.

Type Conversion

Variables must be assigned values that correspond to their type. If an expression involves a mixture of types or if the result is to be placed in a variable of a different type, then it is good practice to force the necessary type conversion explicitly by using the following functions:

Function	Argument Type	Result Type
int	x	i
real	x	r
dble	x	d
ichar	c	i
char	i	c

Here, "c" stands for character, "r" stands for real, "d" stands for double precision, "i" stands for integer, and "x" stands for real, double precision, or integer.

For example, if ksum is an integer variable and sum is real then

$$\text{sum} = \text{real}(\text{ksum})$$

converts the integer representation of ksum to a floating point representation (real), and stores the result in sum. The content of ksum is unchanged.

Similarly, if x is a real variable then
$$i = int(x)$$

converts the floating point number in x to integer format by truncating the fractional part of x and storing the result in integer i. (3.1 becomes 3, -3.1 becomes -3, etc.) The content of x remains unchanged.

Relying on the automatic type-conversion rules of F77 can lead to unexpected results. For example, it is very easy to forget that integer divides produce integer quotients. If i and j are integers and x is real then

$$x = (i/j)*(j/i)$$

has value zero for any unequal nonzero values of i and j.

Another common mistake is to have a stray single precision calculation buried inside a double precision computation. If x and y are real and z is double precision, then

$$z = dble(x)*dble(y)$$

stores the double precision product of x and y in z. Note that this is different from

$$z = dble(x*y)$$

Some Comments about Exponentiation

An exception to the rule against mixing modes concerns exponentiation. If i is an integer and a and b are floating point, then

$$c = a**i + b**i$$

is preferable to

$$c = a**float(i) + b**float(i)$$

This is because in the first assignment, the result is obtained by repeated multiplication, whereas in the second assignment it is computed using logarithms:

```
c = exp(float(i)*log(a))+exp(float(i)*log(b))
```

Another tip that pertains to exponentiation concerns the square root function. If x is a real variable, then it is always better to use `sqrt(x)` instead of `x**.5` .

Specifying Constants in an Expression

Some care must be exercised if a constant appears in an expression. For example, if F and C are double precision then

$$F = (9/5)*C + 32$$

is equivalent to

$$F = C + 32$$

because the "9" and "5" are treated as integers and the integer quotient 9/5 yields 1. Double precision constants should be specified using the "d" notation. For example, 9.0d0, 90.d-1, and .9d1 each specify a double precision version of "9." Thus, for correct double precision centigrade-to-Fahrenheit conversion one should use

$$F = (9.0d0/5.0d0)*C + 32.0d0$$

or

$$F = 1.6d0*C + 32.0d0$$

Real, single precision constants can be written analogously. Just replace the "d" with an "e":

$$F = (9.0e0/5.0e0)*C + 32.0e0$$

This is equivalent to

$$F = (9./5.)*C + 32$$

Not using the "d" and "e" specification format makes a program vulnerable to type errors. It is good practice always to use the "d" and "e" notation when specifying constants.

The "parameter" Statement

Rather than have constants appear in expressions, it is sometimes desirable from the standpoint of readability to define constants with the **parameter** statement. For example, the surface area program at the beginning of this section can be rewritten as follows:

```
      program  area
      real  r,  area,  fourpi
      parameter  (fourpi = 12.566e0)
c
c     This program reads in a real number r and
c     prints the surface area of a sphere having
c     radius  r .
c
      read(*,*)r
      area = fourpi*r*r
      write(*,*)area
      stop
      end
```

The **parameter** statement has the form

parameter ({ *name* = *constant* },..., { *name* = *constant* })

e.g.,

```
      parameter( pi = 3.14159, e = 2.718e0 )
```

Here are the rules associated with the **parameter** statement:

- If a variable appears in a **parameter** statement its value can never change. Thus, a parameter can never be on the left side of the "=" in an assignment statement.

- A variable can appear in at most one **parameter** statement.

- The **parameter** statement(s) in a program must come before the first executable statements.

A convention favored by many writers of large codes is to use symbolic names defined in **parameter** statements for commonly occurring constants, e.g.,

```
parameter ( zero = 0.0e0, one = 1.0e0 )
```

This helps reduce typographical errors and makes it easier to convert a code from one precision to another.

Problems for Section 1.1

1. Complete the following program so that it performs as advertised.

```
      program convrt
      real theta, pi, sixty, s
      integer d, m
C
C     Let theta be the size  of an angle in radians. This pro-
C     gram prints integers d and m  and real s so that
C     theta equals  d degrees + m minutes + s seconds
C
      parameter (pi = 3.14159e0 , c1 = 60.e0, c2 = 180.e0)
      read(*,*)theta
      d =
      m =
      s =
      write(*,*)d, m, s
      stop
      end
```

Do not declare any additional variables. Recall that π radians equals $180°$ and 3600 seconds $= 60$ minutes $= 1°$.

2. Assume that `latA`, `longA`, `latB`, and `longB` are initialized real variables that contain the latitude and longitude of two cities A and B on the surface of the earth, which we assume to be a sphere with radius $R = 3950$ miles. Assume that the latitudes and longitudes are in degrees and that the usual conventions apply. (Latitudes range from -90° (South Pole) to + 90° (North Pole) and the longitudes range from -180° to +180°.) The great circle distance between A and B is the distance of the shortest possible path (on the earth's surface) between A and B. (The path resides on a circle of radius R whose center is the earth's center.) Write a program segment that computes the great circle distance between A and B. Consult Appendix 1.

3. Write a program that reads numbers a, b, c, d, e, f and prints the area of the triangle having vertices (a,b), (c,d), and (e,f). All variables and computations should be double precision.

1.2 LOGICAL OPERATIONS

We now show how to test conditions and use the various **if** constructs that are available in F77.

Logical Variables

Just as numeric values can be stored in real, integer, and double precision variables, truth values can be stored in variables that have been typed "logical." Thus, the declaration

```
logical a, b
```

establishes two variables, each of which may be assigned either the value .TRUE. or the value .FALSE. . The periods are part of the syntax and so, for example,

```
a = .TRUE.
```

is the correct way to assign .TRUE. to a.

Logical Expressions

The truth value of a logical expression can be assigned to a logical variable. A logical expression is formed by comparing arithmetic expressions with the following *relational operators* :

.LT.	less than
.LE.	less than or equal
.EQ.	equal
.NE.	not equal
.GT.	greater than
.GE.	greater than or equal

For example, if x, y, and z are integer variables then

$$a = x \ .EQ. (y + z)$$

is .TRUE. if and only if the contents of y and z when added exactly equal the value of x.

Logical expressions can be combined using the logical operators AND, OR, and NOT. These operators are defined by the following truth tables:

a	b	a .AND. b
.FALSE.	.FALSE.	.FALSE.
.FALSE.	.TRUE.	.FALSE.
.TRUE.	.FALSE.	.FALSE.
.TRUE.	.TRUE.	.TRUE.

a	b	a .OR. b
.FALSE.	.FALSE.	.FALSE.
.FALSE.	.TRUE.	.TRUE.
.TRUE.	.FALSE.	.TRUE.
.TRUE.	.TRUE.	.TRUE.

a	.NOT. a
.FALSE.	.TRUE.
.TRUE.	.FALSE.

Note the "dot" delimiters.

Examples clarify how these operators can be used. Suppose x is an initialized integer variable. Then

$$a = ((x/2)*2 \ .EQ. \ x \ ) \ .OR. \ ((x/3)*3 \ .EQ. \ x \ )$$

is .FALSE. if and only if neither 2 nor 3 divide the contents of x. On the other hand,

$$a = ((x/2)*2 \ .EQ. \ x \ ) \ .AND. \ ((x/3)*3 \ .EQ. \ x \ )$$

is .TRUE. if and only if the content of x is divisible by 6.

The statement

```
a = 3 .LE. i .LE. 7
```

is illegal because relations compare the value of arithmetic expressions. To test if *i* is between 3 and 7:

```
a = (3 .LE. i) .AND. (i .LE. 7)
```

The order of precedence in a logical expression is important. Arithmetic expressions are evaluated first, then all the relational operators are evaluated, and finally all the logical operators are evaluated. For example

```
a = (i .GT.j) .OR. ((j.GE.0) .AND. (j.LT.10))
```

is .TRUE. if and only if $i > j$ or $0 \le j < 10$. On the other hand,

```
a   =   i .GT.j   .AND.   j.GE.0 .OR.   j.LT.10
```

is .TRUE. if $0 \le j < i$ or if $j < 10$.

Follow these guidelines when performing logical computation:

- Only compare variables that have the same type.

- Use parentheses in complicated logical expressions, as the precedence of the logical operations is not widely known.

"if" Constructs

There are several **if** constructs in F77. The simplest is the logical **if** statement:

if ({ *logical expression* })   { *executable statement* }

For example, the following statement prints the value of count if it is positive:

```
if ( count .GT. 0 )    write(*,*)count
```

If "there is more than one thing to do" when the tested logical expression is true then use the construction

if ({ *logical expression* }) **then**

{ *statements* }

endif

Thus, if the value of `count` is positive then the following prints `count` and decreases it by one:

```
if ( count .GT. 0 )     then
   write(*,*)count
   count = count - 1
endif
```

If there are different things to do depending upon the value of the conditional then use the **if-then-else** construction:

if ({ *logical expression* }) **then**

{ *statements* }

else

{ *statements* }

endif

Thus,

```
if ( count .GT. 0)      then
   write(*,*)count
   count = count - 1
else
   count = 100
endif
```

prints and decrements `count` if it is positive but otherwise sets it to 100.

The most general **if** construct has the form

if ({ *logical expression* }) **then**

{ *statements* }

elseif ({ *logical expression* }) **then**

{ *statements* }

:

elseif ({ *logical expression* }) **then**

{ *statements* }

else ({ *logical expression* })

{ *statements* }

endif

Execution flows from top to bottom. The first true logical expression prompts the execution of the associated conditional code. Control is then transferred to the next statement after the **endif**. The final **else** is optional.

Here is a program that reads three distinct integers and prints the median:

```
program median
integer x,y,z,median
read(*,*)x,y,z
if ( (x-z)*(x-y) .LT. 0 ) then
   median = x
elseif ( (y-x)*(y-z) .LT. 0 ) then
   median = y
else
   median = z
endif
write(*,*)median
stop
end
```

It is also possible to nest **if** constructs. The following program segment prints the values of the variables x, y, and z in descending order, provided x is the largest

```
if ( x .GT. y  .AND. x .GT. z ) then
    if ( y .GT. z) then
        write(*,*)x, y, z
    else
        write(*,*)x, z, y
    endif
endif
```

Stylistic Considerations

We have chosen to capitalize the relational operators for emphasis. A more important stylistic concern involves indentation. Appropriate indentation enhances the readability of **if-then-else** constructs. To appreciate this point consider the unindented version of the preceding program segment:

```
if ( x .GT. y  .AND. x .GT. z) then
if ( y .GT. z) then
write(*,*)x, y, z
else
write(*,*)x, z, y
endif
endif
```

"Straightline" code like this is difficult to read, especially in long programs.

Lastly, you should try to avoid the excessive nesting of **if** statements. Things get hard to follow when there are more than three levels of nesting.

Problems for Section 1.2

1. Assume that x and y are initialized integer variables. Write a program segment that swaps the contents of these two variables (if necessary) so that x .GE. y has value .TRUE. .

2. Assume that x, y, and z are integer variables containing three distinct values. Write a program segment that rearranges the contents of these variables so that

x .GT. y   .AND. y .GT. z

has value .TRUE.. Strive to make your program segment as short as possible. Use as many auxiliary variables as you please.

3. Write a program that reads in a real number x and prints

$$x^3 - 2x^2 + 3x - 5 \qquad \text{if} \qquad 0 \le x \le 1$$
$$x^3 + x^2 - 7x + 6 \qquad \text{if} \qquad 1 < x \le 2$$
$$x^3 - x + 8 \qquad \text{if} \qquad x < 0 \text{ or } x > 2$$

4. Assume that a and b are initialized logical variables. Show how the following program segments can be rewritten without nesting.

```
(a)     if ( a ) then
            if ( b ) then
                i = 1
            else
                i = 2
            endif
        endif

(b)     if ( a ) then
            if ( .NOT. b)   i = 1
        else
            if ( b )   i = 2
        endif
```

1.3 LOOPS

F77 supports one loop: the do-loop. However, it is possible to simulate while- and until-loops using **if** constructs.

While-Loops

A while-loop can be realized as follows:

{ *label* }   **if** ({ *logical expression* }) **then**

{ *statements* }

goto { *label* }
endif

The body of the loop is executed so long as the conditional is true. The **goto** statement passes control to the labeled **if** statement. The label is a number between 1 and 99,999 and must be situated within columns 2 through 5. All the labels in a program must be unique.

To illustrate the while-loop construct consider the generation of the Fibonacci sequence { x_n } defined by the recurrence $x_n = x_{n-1} + x_{n-2}$ with $x_{-1} = x_0 = 1$. The following program prints Fibonacci numbers that are strictly less than 200:

```
      program fibon
      integer  x, y, z
      x = 1
      y = 1
10    if (x .LT. 200)    then
         write(*,*)x
         z = x + y
         y = x
         x = z
         goto 10
      endif
      stop
      end
```

25

Until-Loops

Until loops can also be implemented with the **if** and **goto** constructs:

{ *label* }   **continue**

{ *statements* }

if ({ *logical expression* }) **goto** { *label* }

The following program prints all Fibonacci numbers up to and including the first one strictly larger than 200.

```
        program: fibonacci
        integer  x, y, z
        x = 1
        y = 1
10      continue
           z = x + y
           y = x
           x = z
           write(*,*)x
        if (x .LE. 200)   goto 10
        stop
        end
```

Do-Loops

A while-loop can be used for simple counting. For example,

```
        x = 1
        y = 1
        count = 1
10 if ( count .LE. 10 ) then
           z = x + y
           y = x
           x = z
           write(*,*)x
           count = count + 1
           go to 10
        endif
```

prints the Fibonacci numbers $x_1,...,x_{10}$ However, in such cases it is preferable to use the do-loop construction:

```
     x = 1
     y = 1
     do 10 count = 1,  10
        z = x + y
        y = x
        x = z
        write(*,*)x
  10 continue
```

Counting in a do-loop need not be in steps of one. The following program segment computes $s = n(n-1) \cdots (n-k)$:

```
     s = n
     do 20 j = n-1,  n-k,  -1
        s = s*j
  20 continue
```

The **do** statement here says "step from n-1 to n-k in steps of -1." This example reveals the general form of the do-loop:

> **do** { *label* } { *var* } = { *exp.1* }, { *exp.2* }, { *exp.3* }
>
>   { *statements* }

{ *label* }   **continue**

Here, *var* is the loop index, *exp.1* is an expression that specifies the initial value of *var*, *exp.2* is an expression that specifies the terminating bound, and *exp.3* is an expression that defines the increment. To clarify do-loop termination when the counting does not "land" on the termination value we remark that

```
     do 10 count = i1,  i2,  i3
        { loop body }
  10 continue
```

is equivalent to

```
          count = i1
    10 if (count .LE. i2  )    then
               { loop body }
            count = count + i3
            goto 10
          endif
```

if the value of the increment i3 is positive. If the value of i3 is negative, then the above do-loop template is equivalent to

```
          count = i1
    10 if (count .GE. i2  )    then
               { loop body }
            count = count + i3
            goto 10
          endif
```

Note that the entire loop is skipped if the termination criteria are fulfilled when the **do** statement is initially processed.

Follow these rules when constructing do-loops:

- The do-loop variable, as well as the expressions in the do-loop statement, should have type integer.

- It is permissible to omit the third do-loop parameter, which specifies the increment. When this is the case a unit increment is presumed.

- The do-loop variable must never be changed by statements within the body of the loop.

Nesting Do-Loops

Do-loops may be nested. Here is a program segment that prints all integer triplets (i, j, k) with the property that $1 \leq i \leq j \leq 100$ and $k^2 = i^2 + j^2$.

```
      do 10 i = 1,100
         do 5  j = i,100
            k = int( sqrt( real(i*i + j*j))
            if (k*k .EQ. i*i + j*j) then
               write(*,*)i,j,k
            endif
  5      continue
 10 continue
```

More on "goto" and "continue"

Execution of a Fortran program begins at the top of the main program and flows to the bottom. However, branches to previous or subsequent lines are sometimes needed as in the while-loop construction. Such branches can be effected through the use of the **goto** statement as we have seen. The form of this statement is

$$\text{\textbf{goto} } \{ \textit{ label } \}$$

where { *label* } is a number listed in columns 2 through 5 of the line of the program to which the branch is being made. The "no operation" **continue** statement is useful for specifying the target of the **goto**. For example,

```
      10 continue
```

{ *statements* }

```
      goto 10
```

causes control to return up to statement 10 and continue (downward) from there. The **continue** statement does nothing but "hold" the label. Control then "drops" to the next line.

The **goto** statement should be used only as a last resort since the presence of **goto**'s in a program makes for difficult reading. The while-loop setting is just about the only tolerable use of **goto**. Avoid using the **goto** to effect a quick exit from a do-loop. In such cases it is better to rewrite the do-loop as an appropriate while-loop.

One final reminder: a do-loop must be entered "from the top." Never jump into the body of a do-loop.

Loops, Indentation, and Labels

As we have illustrated in our examples, readability is enhanced with the appropriate indentation of loops. The body of a loop should be uniformly indented to highlight the program structure. Each loop-nesting requires further indentation.

One final stylistic comment about statement labels, i.e., the numbers found in columns 2 through 5. They should be assigned so that the sequence of numbers is increasing, top to bottom. Moreover, it is good practice to leave gaps between the values of consecutive labels so that subsequent program modifications can be easily accommodated, maintaining an increasing sequence of labels.

Problems for Section 1.3

1. Let n and k be integers that satisfy $1 \le k \le n$ and define

$$bi = n(n-1)(n-2) \cdots (n-k+1) / k!$$

This quantity is always an integer. Correct the following program so that it reads n and k and prints bi:

```
       program bi
       integer n, k, bi, j
       bi = 1
       read(*,*)n, k
       do 10 j = 1,k
          bi = bi*((n - j + 1)/j)
   10 continue
       write(*,*)bi
       stop
       end
```

2. Here is Euclid's algorithm for finding the greatest common divisor g of two positive integers a and b :

Step 1. Divide a into b and let r be the remainder.
Step 2. If $r = 0$ then set $g = a$ and quit.
Step 3. Set $a = b$, $b = r$, and go to Step 1.

Implement this algorithm with a while-loop construct.

3. The following program segment computes the $n = 10$ partial sum of the Taylor series for e^x :

```
sum = 1.0
term = 1.0
do 10 n = 1,10
    term = term*(x/real(n))
    sum = sum + term
10 continue
```

Change the segment so that upon completion sum contains the tenth partial sum for **(a)** $sin(x)$, **(b)** $cos(x)$, and **(c)** $cosh(x)$.

4. Write a program that prints the number of integral index pairs (a,b) of the form $1 \leq a < b \leq 100$ with the property that a divides b.

5. Express the following as a while-loop:

```
eps = 1.0
do 10 k = 1,100
    eps = eps/2.0
    epsp1 = eps + 1.0
    if (epsp1 .EQ. eps)    go to 20
10 continue
20 write(*,*)eps
```

1.4 ARRAYS

Many scientific computations require manipulation of matrices and vectors. Fortran stores matrices and vectors in *arrays*. An array is just a contiguous block of memory with an implicit indexing scheme. An array has a data type.

One-dimensional Arrays

The declaration

```
real a(100)
```

identifies a as a one-dimensional real array of size 100. That is, a is comprised of 100 contiguous memory locations, each capable of storing a real floating point number. It is useful to think of the array a as a linear sequence of cells:

a(1)
a(2)
•
•
•
a(100)

The content of cell 1 is denoted by a(1). The content of cell *i* is denoted by a(i).

Similarly, the statement

```
real b(-41:58)
```

identifies b as a one-dimensional real array of size 100, with index range -41 to 58 :

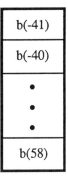

Double precision, integer, and logical arrays are also possible, e.g.,

```
double precision r(100)
integer i(0:99)
logical b(10)
```

In declarations, it is perfectly legal to intermingle simple variables with arrays:

```
real x, y, z(-100:100)
```

A reference to a one-dimensional array in a declaration has the form

{ *name* }({ *first_index* }:{ *last_index* })

The length of the array is thus *last_index* - *first_index* + 1. If *first_index* = 1 then it is legal to dispense with the "colon notation":

```
integer k(100)    ⟺    integer k(1:100)
```

It is important to distinguish between vectors (mathematical objects) and one-dimensional arrays (Fortran objects). A *vector* is stored in an *array*. For example, the following program segment stores the vector (1, 2, 3, 4) in the "middle" of the integer array x.

```
      integer x(10)
      do 5 i=4, 7
         x(i) = i-3
    5 continue
```

Clearly, the size of the vector must be less than or equal to the size of the holding array.

Two-dimensional Arrays

The declaration

```
      integer A(4,5)
```

defines A as a two-dimensional integer array comprised of $4 \times 5 = 20$ integer elements. Double subscripts are used to refer to individual elements of the array:

(1,1)	(1,2)	(1,3)	(1,4)	(1,5)
(2,1)	(2,2)	(2,3)	(2,4)	(2,5)
(3,1)	(3,2)	(3,3)	(3,4)	(3,5)
(4,1)	(4,2)	(4,3)	(4,4)	(4,5)

Thus, if the entire array is initialized then

```
       s = 0
       do 10 j = 1,5
          do 5 i = 1,4
             s = s + A(i,j)
    5     continue
   10 continue
```

computes the sum of the entries. In general, a reference to a two-dimensional array in a declaration has the form

{*name*}({ *first_index1* }:{ *last_index1* },{ *first_index2* }:{ *last_index2* })

The total size L of the array is thus

$$L = (last\_index1 - first\_index1 + 1)\ (last\_index2 - first\_index2 + 1)$$

Again, if an index range is from "1 to something" you can avoid the colon notation:

```
integer k(20,100)    ⟺    integer k(1:20,1:100)
```

Two-dimensional arrays are usually used to store their corresponding mathematical object, matrices. For example, the 3-by-3 "times table" matrix

$$A = \begin{bmatrix} 1 & 2 & 3 \\ 2 & 4 & 6 \\ 3 & 6 & 9 \end{bmatrix}$$

can be stored in the array A as follows:

```
      do 20 j=1, 3
         do 10 i=1,3
            A(i,j) = i*j
10       continue
20 continue
```

It is not necessary for the size of the matrix to match the size of the array but, of course, the matrix must fit into the array. In the example above, the row and column dimensions of the matrix A are less than or equal to the size of the row and column dimensions of the array A . The matrix sits in the "northwest" corner of the array:

1	2	3		
2	4	6		
3	6	9		

It is perfectly legal to refer to an element outside the matrix bounds but within the array bounds, e.g., `val = A(4,5)`. The variable `val` is merely assigned the contents of location (4,5) of the array. Depending on the context, this may be a programming error. Array entries `A(i,j)` may not be initialized if either `i` or `j` is greater than 3. It is entirely the programmer's responsibility to ensure that the index values at execution time are meaningful. Do not assume that array entries are automatically initialized to zero by the compiler.

Space Allocation for Two-dimensional Arrays

Fortran stores a two-dimensional array as a contiguous, linear sequence of cells, by *column*. For example, space allocation for the 4-by-5 array `A` above would be as follows:

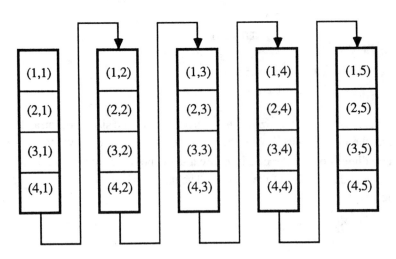

Note that storage of the 3-by-3 times table matrix in the 4-by-5 array results in unused array space:

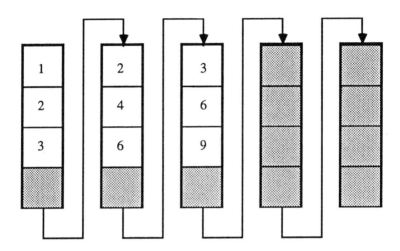

Moreover, the matrix elements are not contiguous in memory.

It is instructive to understand how the "address" of an array element is determined. Suppose that A is an m-by-n matrix stored in array A that has row dimension $adim$. The address of the cell holding the (i, j) entry of the matrix A is given by

$$addr[\,\text{A(i,j)}\,] = addr[\,\text{A(1,1)}\,] + \text{(j-1)*adim} + \text{(i-1)}$$

A typical mistake is to assume that the address is specified by

$$addr[\,\text{A(i,j)}\,] = addr[\,\text{A(1,1)}\,] + \text{(j-1)*m} + \text{(i-1)}$$

This is because space is allocated column by column for the *array* A, not the matrix A. It is interesting to note that neither the column dimension of the array nor the dimensions of the stored matrix figure in the address computation. (Note that "physical" address computation requires knowledge of type as well. For example, a cell may be 4 bytes long for integer arrays and 8 bytes long for double precision arrays.)

In many programs it is the case that the declared arrays are exactly the same size as the matrices they house. Thus, the array/matrix distinction is immaterial. But there are occasions when several matrix problems of unequal size need to be performed and it is convenient to utilize a single array that is just large enough to handle the largest problem. The following example shows how we would go about printing the matrices F_2, F_4, and F_6 and their transposes, where F_n is the n-by-n Frank matrix defined by

$$F_n = (f_{ij}) \qquad f_{ij} = \begin{cases} n - j + 1 & \text{if} & i \leq j \\ n - j & \text{if} & i = j + 1 \\ 0 & \text{if} & i \geq j + 2 \end{cases}$$

Example 1.4-1

```
         program frank
         integer F(15,15), n, i, j
         do 20 n = 2,6,2
            do 10 i = 1,n
               do 5 j = 1,n
                  if (i .LE. j ) then
                     F(i,j) = n-j+1
                  elseif (i .EQ. j+1 ) then
                     F(i,j) = n-j
                  else
                     F(i,j) = 0
                  endif
5                 continue
10             continue
            do 15 i = 1,n
               write(*,*) (F(i,j),j=1,n)
15          continue
            do 20 i = 1,n
               write(*,*) (F(j,i),j=1,n)
20          continue
25       continue
         stop
         end
```

The **write** involves a new construct that is handy when printing array elements. When $n =$ 4, for example,

```
write(*,*) (F(i,j),j=1,n)
```

is equivalent to

```
write(*,*)F(i,1),F(i,2),F(i,3),F(i,4),F(i,5)
```

Because matrices can be large, an entire matrix row or column may not fit on a single output record. The resulting output may not be particularly attractive and it may be necessary to employ the formatted **write** construct. This is discussed in §1.6.

Packed Storage

Matrices are two-dimensional and therefore it is natural to store them in their F77 counterpart, two-dimensional arrays. As we have mentioned, a two-dimensional array is physically stored as a long one-dimensional sequence with an implicit two-dimensional indexing scheme. However, there are occasions when it is more efficient to explicitly store a matrix in a one-dimensional array and to simulate the two-dimensional indexing. This occurs if the matrix has a lot of zeros or some regular, exploitable structure .

As an example, let us return to the times table matrix. This matrix is symmetric and so it is redundant to store both a_{ij} and a_{ji}. Suppose we store only the lower triangular portion in column-by-column fashion in a one-dimensional array. Thus, in the $n = 4$ case we have the following identification:

$$
A = \begin{bmatrix} 1 & 2 & 3 & 4 \\ 2 & 4 & 6 & 8 \\ 3 & 6 & 9 & 12 \\ 4 & 8 & 12 & 16 \end{bmatrix} \quad \Leftrightarrow \quad a = \begin{bmatrix} 1 \\ 2 \\ 3 \\ 4 \\ 4 \\ 6 \\ 8 \\ 9 \\ 12 \\ 16 \end{bmatrix}
$$

By storing *a* instead of *A* we approximately halve the amount of required storage. In this case we say that the matrix is stored in *packed* form. It is a useful data structure when dealing with symmetric and triangular matrices.

Unfortunately, if matrices are stored in packed form the programmer must look after the two-dimensional indexing explicitly. For example, to set up an n-by-n times table matrix in packed form a one-dimensional array a of length at least $n(n+1)/2$ must be declared. The setting up then takes the form:

```
      k = 1
      do 10 j = 1,n
         do 5 i = j,n
c
c     a(k) = A(i,j)
c
            a(k) = i*j
            k = k+1
    5    continue
   10 continue
```

Note the need for the "hand-indexing" variable k .Writing clear programs that manipulate matrices that are stored in the packed format is a challenge. Comments should be used to explain the program subscripting.

Arrays of Higher Dimension

It is possible to manipulate arrays of dimension up to seven. For example, the declaration

```
      real A( 2, 3, 6, 2, 8, 7, 4 )
```

sets aside enough memory for a seven-dimensional array. Note that $2\times3\times6\times2\times8\times7\times4 = 16,128$ memory locations are reserved. High-dimensional arrays get big quickly.

As an example of a multidimensional array problem, suppose hcube is a 2-by-2-by-2-by-2 integer array. Such an array would be declared as follows:

```
      integer hcube( 0:1,0:1,0:1,0:1,0:1 )
```

Suppose we now want to assign $(abcd)_2$ to *hcube(a,b,c,d)*. This can be accomplished as follows:

```
      do 40 a = 0,1
         do 30 b = 0,1
            do 20 c = 0,1
               do 10 d = 0,1
                  hcube(a,b,c,d) = 8*a+4*b+2*c+d
10               continue
20            continue
30         continue
40 continue
```

We mention that if A is a d-dimensional array with component dimensions $n_1,...,n_d$ and $N_k = n_1 \ldots n_k$ then the "address" of $A(i_1,...,i_d)$ is given by

$$addr(i_1,...,i_d) = (i_d - 1)N_{d-1} + (i_{d-1} - 1)N_{d-2} + \cdots + (i_1 - 1)N_0 + 1$$

Problems for Section 1.4

1. Supppose n-by-n upper triangular matrices A and B are stored in n-by-n arrays A and B. Write a program segment that overwrites A with the product AB. Try not to use any auxiliary storage.

2. Suppose that the array A(0:101,0:101) has been initialized. Write a program segment that overwrites A(i,j) with the average of its four "neighbors" A(i-1,j), A(i+1,j), A(i,j-1), and A(i,j+1) where we assume that $1 \le i \le 100$ and $1 \le j \le 100$. Try to get by with a minimum amount of auxiliary storage.

3. Let E be the n-by-n down-shift matrix, i.e., $w = Ev \Rightarrow w_i = v_{i-1}$ for $i = 2{:}n$ and $w_1 = v_n$. Write a program segment that sets up the n-by-n matrix A whose kth column equals $E^{k-1}v$, where v is a given n-vector.

4. The n-by-n Pascal matrix A_n has 1's in the first column and row, and "interior" entries prescribed by $a_{ij} = a_{i-1,j} + a_{i,j-1}$. Write a program that sets up A_n. What is the largest n for which A_n can be precisely stored in your computer?

1.5 SUBPROGRAMS

Subprograms are used to encapsulate important computations. A computation may be important because it is repeatedly performed or because it has a logical structure that is worth separating from the main program.

In this section we cover the two types of subprograms that are allowed in F77: functions and subroutines. There are special issues associated with the use of arrays and subprograms and they are the subject of §1.6.

Built-in Functions

It is useful to preface the discussion of "user-supplied" subprograms with some remarks about the built-in or "intrinsic" functions that are a part of F77 itself. A summary of these functions is given in Appendix 1.

It is necessary to distinguish between two kinds of built-in functions in F77: *specific* and *generic*. A specific function has a data type (e.g., real, double precision, etc.). For example, the integer-to-character conversion function **char** has data type character because it returns a string. When specific functions are used they should appear in the type declaration statements.

A *generic* intrinsic function does not have a data type *per se*. The data type of the returned value of a generic intrinsic function is determined by the data type of its arguments. Thus, if the statement

```
x = cos(y)
```

appears in a program, then x and y should be of the same type because cos returns a real value if y is real and a double precision value if y is double precision. A generic built-in function should *not* be declared in a type declaration statement.

Built-in functions can appear in assignment statements. Thus, if pi is double precision then

```
pi = 4.0d0*atan(1.0d0)
```

computes a double precision version of π while

```
    s = cos(theta)*sin(phi) + cos(phi)*sin(theta)
```
and

```
    s = sin(theta + phi)
```

are two different ways to compute $sin(\theta + \varphi)$.

Recognize that certain functions place restrictions on the value of the argument and that fatal errors can result when these restrictions are violated, e.g., `sqrt(-1)`.

Some built-in functions accept more than one argument. Thus,

```
    a = max(sx1,sx2)
```

assigns the larger of the two values in `sx1` and `sx2`, whereas

```
    b = min(sx1,sx2)
```

assigns the smaller value. In these examples, a, b, `sx1`, and `sx2` are real.

See Appendix 1 for a list of available built-in functions and for a clarification of the generic versus specific issue.

Functions

We now consider user-written subprograms. To impose a continuity on the discussion, many of our examples revolve around the following mathematical problem:

Problem P : Given a function $f:R \rightarrow R$ and a positive integer n, compute $m(f,n) = min \{ f(0) ,..., f(n) \}$.

Thus, if $n = 2$ and $f(x) = x^2 - 2x + 3$, then $m(f,n) = 2$.

Suppose that we are to solve problem **P** for the specific case,

$$n = 10$$

$$f(x) = x^3 - 2x^2 - 7x - 3$$

This can be done using existing techniques:

```
        program min1
        integer k
        real m
c
c   Set m = f(0)
c
        m = -3.0e0
c
c   Now check f(1),...,f(10)
c
        do 10 k = 1,10
           x = real(k)
           m = min(m,((x - 2.0)*x - 7.0)*x - 3.0)
     10 continue
        write(*,*)m
        stop
        end
```

However, greater clarity can be achieved if we define a **function** f ,

```
        real function f(x)
        real x
        f =  ((x - 2.0)*x - 7.0)*x - 3.0
        return
        end
```

and then execute the main routine:

```
        program min2
        integer k
        real m, f
        m = f(0.0e0)
        do 10 k = 1,10
           m = min( m , f( real(k) ) )
     10 continue
        write(*,*)m
        stop
        end
```

As seen by the example, a **function** in F77 has the structure

> {*type* } **function** { *name* } ({ *list-of-variables* })
> { *declarations* }
> { *statements* }
> **return**
> **end**

Here are some rules associated with user-defined functions:

- The value produced by the function is returned to the calling
  program via a variable whose name is the function's name, e.g.,
  f(x)   is returned in the variable f.

- A function has a type and it must be declared in the calling program.

- The body of a function resembles the body of a main program and
  the same rules apply.

- When a function is called, execution begins at the top and flows
  to the bottom. When a **return** statement is encountered the subroutine
  is exited and control is passed back to the calling program.

- A reference to a defined function can appear anywhere in an
  expression, subject to type considerations.

Functions with Other Functions as Arguments

Now suppose that we have two functions $f_1(x)$ and $f_2(x)$ and that
we wish to compute $m(f_1,10)$ and $m(f_2,20)$. An obvious way to solve this
problem would be to expand the main program:

```
program min3
integer k
real m1, m2
m1 = f1(0.0e0)
```

```
      do 10 k = 1,10
         m1 = min( m1, f1( real(k) ) )
10    continue
      m2 = f2(0.0d0)
      do 20 k = 1,20
         m1 = min( m1 , f1( real(k) ) )
20    continue
      write(*,*)'m(f1,10)=',m1,'m(f2,20)=',m2
      stop
      end
```

However, a better solution is to define the function

```
      real function fmin(f,n)
      integer n
      real f
      fmin = f(0.0d0)
      do 10 k = 1,n
         fmin = min( fmin ,f(real(k)) )
10    continue
      return
      end
```

and then execute the more readable main routine

```
      program min4
      real m1, m2, f1, f2
      external f1, f2
      m1 = fmin( f1,10 )
      m2 = fmin( f2, 20 )
      write(*,*)'m(f1,10)=',m1,'m(f2,20)=',m2
      stop
      end
```

The example introduces the **external** statement. Any time a function or a main program passes the name of a function to another function it must identify the name in an **external** statement:

external { *list of function names* }

The **external** statement should be placed before any executable code. **External** statements are necessary because subprograms are compiled independently and there is no other way to inform the compiler that a particular argument names a subprogram.

Common

Now suppose that we want to solve problem **P** for arbitrary n and arbitrary cubics $f(x) = ax^3 + bx^2 + cx + d$. The function

```
real function f(x,a,b,c,d)
real x,a,b,c,d
f = ((a*x + b)*x + c)*x + d
return
end
```

is a perfectly legal way to evaluate f but it has the flaw that we can no longer use the function fmin to compute $m(f,n)$. The reason is that fmin expects a function with a single argument, i.e., f(x), not a function of the form f(x,a,b,c,d). One way around this is to modify fmin so that it accepts functions with five real arguments. However, if we think of fmin as a general routine for solving problem **P** then this is not an acceptable solution.

An alternative solution involves a new construct--the **common** statement. By placing variables "in common" they can be shared between the main program and one or more subprograms. We can place the variables a, b, c, and d in common as follows:

```
program min5
integer n
real m, a, b, c, d, f
common /coeff/ a, b, c, d
read (*,*)n, a, b, c, d
m = fmin(f,n)
write(*,*)m
stop
end
```

This program prints $m(f,n)$, provided we identify the common variables in the function f as well:

```
real function f(x)
real x
real a,b,c,d
common /coeff/ a, b, c, d
f = ((a*x + b)*x + c)*x + d
return
end
```

Prompted by this example, we draw some conclusions.

- The syntax for a **common** statement is as follows:

  common / { *name* } / { *list-of-variables* }

  Different common blocks must have different names.

- It is legal for a variable to belong to more than one common block.

- The variables listed in a common block are shared by all subprograms that list the block.

- The **common** statement should be placed before any executable statements. The variables listed in a common block must be declared in every subprogram using the common block.

- The common variables do not have to be named the same in each such routine, but they must be declared to have the same type and be listed in the same order.

As an illustration of this last guideline, if

```
integer j, k
real x, y, z
common /eg/ x, y, z, j, k
```

appears in the main program then

```
integer m,n
real a,b,c
common /eg/ a,  b,  c,  m,  n
```

may legally appear in a subprogram and

```
integer m,n
real a,b,c
common /eg/m,n,a,b,c
```

results in a type error. A subprogram may use only part of a common block. Thus,

```
integer m
real a,b,c
common /eg/a,b,c,m
```

is legal. However, it is good practice not to take such shortcuts when using **common**.

Subroutines

Let us return to problem **P**. Suppose that in addition to computing $m(f,n)$ we want to determine an integer $k \in \{1,...,n\}$ so $f(k) = m(f,n)$, i.e., the minimizing point. Since a Fortran function can essentially return only one value, we cannot readily approach this new problem by modifying the function fmin. A different kind of subprogram called a **subroutine** should be used. Here is an example that solves the problem of computing both the minimum value and the minimizing point:

```
subroutine fmin( f, n, value, point)
integer n, point
real f, value
integer i
real temp
```

```
      point = 0
      value = f(0.0e0)
      do 10 i = 1,n
         temp = f(real(i))
         if (temp .LT. value) then
            point  = i
            value = temp
         endif
   10 continue
      return
      end
```

The subroutine `fmin` can then be "called" from the main routine:

```
      program min6
      external f
      integer n, mpt
      real mval
      read(*,*)n
      call fmin( f, n, mval, mpt )
      write(*,*)'m(f,n)=',mval, 'mpt = ', mpt
      stop
      end
```

Execution of the main program `min6` results in the printing of $m(f,n)$ and the minimizing point.

In general, subroutines are structured as follows:

> **subroutine** {name} ({ *argument list* })
> **declarations**
>
> { *statements*}
>
> **end**

Here are some rules associated with subroutine use:

- A subroutine's arguments are used to exchange information between the subroutine and the calling program. Typically, some of the arguments provide input to the subroutine, e.g., `f, n`, while other arguments are used to convey the results back to the calling program, e.g., `value, point`.

- A subroutine is called by a statement of the form

  call { *name* } ({ *argument list* })

- The arguments in the **call** must agree in number and type with the arguments specifed in the **subroutine** statement. The *names* of the listed variables in the **call** need not agree with the names of the "dummy" variables used in the subroutine definition. In the above example, `mval` and `mpt` are identified with the dummy variables `value` and `point`, respectively.

- When a subroutine is called, execution begins at the top and flows to the bottom. When a **return** is encountered control passes back to the line after the **call** in the calling program.

- When calling a subroutine with an argument list, the Fortran 77 compiler actually passes the addresses of the arguments as opposed to the values themselves. Usually this does not affect the programmer one way or the other, but it does play an important role when passing arrays where the address of the first element is passed. We explore this in more detail in §1.6.

- Unlike functions, subroutines do not have a type. The name of a subroutine does not return a value.

Functions versus Subroutines

Every function can be put into "equivalent" subroutine form. For example the following program prints $f(x)$, where $f(x) = ax^3 + bx^2 + cx + d$:

```
program print
real a, b, c, d, x0, value, f
read(*,*) a, b, c, d, x0
value = f(a, b, c, d, x0)
write(*,*) value
stop
end

real function f(a, b, c, d, x)
real a, b, c, d, x
f = ((a*x + b)*x + c)*x + d
return
end
```

In "subroutine language" this becomes

```
program print
real a, b, c, d, x0, value
read(*,*)a, b, c, d, x0
call f(a, b, c, d, x0, value)
write(*,*) value
stop
end

subroutine f(a, b, c, d, x, value)
real a, b, c, d, x, value
value =   ((a*x + b)*x + c)*x + d
return
end
```

In general, it is better to cast a subprogram in function form whenever a single value is returned. This is because function references can appear in expressions and the resulting code has more the appearance of ordinary mathematics.

We point out that it is not strictly correct to say that a function can return only one value, insofar as a function can modify its arguments just like a subroutine. For example,

```
real function f(x,n)
real x
integer n
```

```
      f = exp(x)
      n = n+1
      return
      end
```

returns $f = e^x$ and increments n. We do not recommend using functions in this capacity. Subroutines should be used whenever a subprogram must report more than one value.

Save

Ordinarily, the local variable values in a subprogram are lost when control passes back to the calling program. However, it is possible to retain the value of a local variable if it is named in a **save** statement. Here is an example:

```
      subroutine print(k)
      integer k
      integer lastk
      save lastk
c
c     This subroutine prints the value of k if it
c     is zero or different from the last printed
c     value. The first call must be with k = 0.
c
      if (k .EQ. 0 ) then
         write(*,*)k
         lastk = k
      elseif ( k .NE. lastk ) then
         write(*,*) k
         lastk = k
      endif
      return
      end
```

The **save** statement should be placed before the first executable statement in the subprogram and it should have the form

$$\textbf{save} \ \{ \ \textit{list-of-variables} \ \}$$

Further Rules and Guidelines

There are a few aspects about subprograms that we have overlooked or not stressed enough:

- Subprogram names are subject to the same rules as variable names.

- Local variables exist only in the subprogram. They do not retain their values when the subprogram is left and then re-entered unless they are named in a **save** statement.

- Subprograms can be invoked by other subprograms as well as by the main program.

- All functions used by a subprogram should be declared. This includes all *specific* built-in functions.

- To enhance readability, the logic of a subprogram should be arranged so that there is only one **return.**

- Minimize the use of **common.** It couples otherwise distinct logical units in a program and complicates readability.

Subprogram Style

In the interest of space we have suppressed comments in most of our examples in this section. Of course, in practice it is very important that subprograms be amply commented so that their use is straightforward. Other guidelines should be followed when writing subprograms and we express these in the form of a template. (For functions, replace the first line with " {type} **function** ({ argument list }) " .)

```
        subroutine { name } ( { argument list } )
c
            { argument declarations and comments}
c
            {Comments about the mission of the subprogram including a
```

```
                  clear specification of subprogram input and output}
      c
      c   Common blocks
      c
                  { common blocks  and comments}
      c
      c   Local Variables and Built-in Specific Functions
      c
                  { declarations and comments }
      c
                  { external statements and comments }
      c
                  { parameter statements and comments }
      c
                  { program body  and comments }
      c
                  return
                  end
```

Of course, some parts of this template may be missing in a particular situation. For example, there may be no common blocks.

To illustrate these stylistic considerations we rewrite the programs min5, fmin, and f above with a full complement of comments .

```
            program min5a
            integer n
            real m, a, b, c, d, fmax
      c
      c   This program prints the minimum value of the
      c   cubic f(x) = ((ax + b)x + c)x + d on the set
      c   {1,2,...,n}.
      c
      c       Uses fmax.
      c
      c       Common blocks
      c
            common /coeff/ a, b, c, d
      c
      c       Read n and the coefficients a,b,c, and d.
      c
```

```
      read (*,*)n, a, b, c, d
c
c  Compute the maximum and print.
c
      m = fmax(f,n)
      write(*,*)m
      stop
      end

      real function fmin(f,n)
c
      integer n
      real f
c
c  This function returns m(f,n) =
c  min { f(0) ,..., f(n) }, where f is a real
c  function of a single argument.
c
c  Local Variables and Intrinsic functions
c
      integer k
      real f
c
      fmin = f(0.0e0)
      do 10 k = 1,n
         fmin = min( fmin , f( real(k) ) )
   10 continue
      return
      end

      real function f(x)
      real x
c
c  This function returns the value of the
c  cubic f(x) = ((ax + b)x + c)x + d.
c
      real a, b, c, d
      common /coeff/ a, b, c, d
      f = ((a*x + b)*x + c)*x + d
      return
      end
```

Problems for Section 1.5

1. Write a function h(f,g,x) that returns $h(x) = f(g(x))$, where f and g are real-valued functions that are defined everywhere.

2. Assume that the subroutine mini(f, a, b) returns the minimum value of the real-valued function $f(x)$ on $[a,b]$. How could this subroutine be used to compute

$$\min_{\substack{i \in \{1,...,100\} \\ 1 \le x \le 100}} x^3 - 36ix^2 + 7i^2x - 9i^3$$

3. Write a function dist(lata, longa, latb, longb) that returns the straight-line "tunnel" distance between two points A and B on the surface of the earth whose latitude and longitude coordinates (lata,longa) and (latb, longb) are given in degrees. Assume that the earth is a perfect sphere of radius 4000 miles and that the latitudes and longitudes are given in degrees:

$$-90° \text{ (South Pole)} \le \text{latitude} \le +90° \text{ (North Pole)}$$

$$-180° \text{ (East of Greenwich)} \le \text{longitude} \le +180° \text{ (West of Greenwich)}$$

4. Building upon Problem 3, write a subroutine

```
dist2(lata,longa,latb,longb,tunnel,gcircle)
```

that returns both the tunnel distance and the great circle distance between A and B. The great circle through A and B is the circle of radius 4000 that lies in the plane of A, B, and C (earth center) and has its center at C. The great circle distance is the length of the shorter of the two arcs defined by A and B on the great circle.

1.6 ARRAYS AND SUBPROGRAMS

A local array can be defined within a subprogram. For example, in

```
subroutine mult(H,x)
double precision H(5,5)
     :
```

the double precision two-dimensional array H is defined and can be used within this subroutine. However, it is not good programming practice to use local arrays except in cases where the array size is small and fixed. Instead, it is preferable to define each array in the main program and then to pass it and the relevant dimension information as arguments. This allows for more flexible subprograms since all concerns regarding space allocation, as well as input and ouput of arrays, are dealt with in the main program. Most scientific software packages adhere to this policy of array use and so it is important that it be understood.

Subprograms with Array Arguments

In Fortran, an array that is defined in the main program and then passed to a subprogram as an argument does not physically move. Instead, the *address* of the "top" of the array is passed. The compiler can then "figure out" all the necessary addresses whenever an element of the array is accessed. To illustrate this point consider the following subroutine that performs matrix-vector multiplication

```
      subroutine matvec( p, q, C, cdim, v, w )
      integer p, q, cdim
      real v(*), w(*), C(cdim,*)
c
c  Compute w := Cv where C is a p-by-q matrix
c  contained in the array C that has row dimension
c  cdim.
c
c  Local variables
      integer i,j
```

```
c
      do 10 i = 1, p
         w(i) = 0.0e0
   10 continue
      do 30 j = 1, q
         do 20 i = 1, p
            w(i) = w(i) + C(i,j)*v(j)
   20    continue
   30 continue
      return
      end
```

Notice that matvec requires the row and column dimensions of the matrix C and the row dimension of the array C.

A main program that calls matvec could be structured as follows:

```
      integer idim, jdim
      parameter (idim = 50, jdim = 40)
      integer i, j, m, n
      real A(idim, jdim), x(jdim), y(idim)
         :
c
c     Compute y = Ax
c
      call matvec(m, n, A, idim, x, y)
         :
      stop
      end
```

This example illustrates a number of key ideas. It is fairly obvious that the row and column dimensions of the *matrix A* must be passed in m and n for otherwise it would not be identifiable. However, it is also essential that idim, the row dimension of the array, be passed to matvec as well. When matvec is called the address of A(1,1) is passed. Within matvec, the location of element (i,j) of the matrix C is computed as we discussed in §1.4:

$$addr[\, C(i,j)\,] = addr[\, C(1,1)\,] + (j-1)*cdim + (i-1)$$

Thus, if A is identified with C then idim must be identified with cdim.

That is why the row dimension of the array A must be passed to matvec.

Another lesson to be learned from the matrix-vector multiply example pertains to the use of the asterisk in the type declaration statement in subprograms. The F77 compiler knows that v and w are to be treated as one-dimensional arrays because they are both listed with one argument in the type declaration statement. The asterisk acts as a "placeholder." Likewise the compiler knows that C is to be treated as a two-dimensional array because it is listed with two arguments in the type declaration statement. The number of arguments listed indicates the assumed dimension of the array in the subprogram. The last dimension of an F77 array is not needed for address computation, so asterisks suffice.

The asterisk could be replaced with any positive integer without harm since it is used only as a marker. In fact, the old Fortran did not allow asterisks and an argument of "1" was commonly used.

Different Dimensions

If a variable is used as a two-dimensional array in the body of a subroutine, then it must be declared as a two-dimensional array in the subroutine's type declaration statement. In the matvec example, C is used as a two-dimensional array in the body of matvec because it appears with double indices in the declaration.

However, the dimension of a passed array does not have to conform to its dimension in the calling program. It is perfectly legal to pass a two-dimensional array to a subprogram and then to treat it as one-dimensional in the subprogram, provided care is exercised. For example, suppose that we have at our disposal the function

```
      integer function prod(n,x,y)
      integer n, x(*), y(*)
c
c     prod = x^Ty where x, y are integer n-vectors
c
      integer i
      prod = 0.
      do 10 i = 1, n
         prod = prod + x(i)*y(i)
 10   continue
      return
      end
```

Assume that A is an m-by-n integer matrix stored in a two-dimensional integer array A, whose row dimension is stored in adim. Assume that m and n contain A's dimensions and that we want to compute

$$frob2(A) \;=\; \sum_{j=1}^{n} \sum_{i=1}^{m} a_{ij}^{2}$$

Observe that $frob2(A) = v^{T}v$, where v is obtained by stacking the columns of the matrix A on top of one another. Mindful that F77 stores arrays by column, it is tempting to compute $frob2(A)$ as follows:

```
length = m*n
frob2A = prod( length , A, A )
```

Note that the two-dimensional array A is treated as a one-dimensional array in prod. This is fine because only the address of the top of the array A is being passed. However, this solution is *not correct* unless m and adim have the same value. This is because the entries of the matrix A are not contiguous unless the row dimensions of the matrix and the holding array are the same. A better way to compute frob2 (A) would be one column at a time:

```
frob2A = 0.
do 10 j = 1,n
    frob2A = frob2A + prod( m,A(1,j),A(1,j) )
10 continue
```

Note that the address at the top of the jth column is passed at iteration j. As far as prod is concerned this is just the address of the top of the vectors x and y. Finally, note that it is not necessary for prod to "know" adim because only single, designated columns of A are referenced during a given call.

Passing Submatrices

Suppose A is an m-by-n matrix and that the integers i_1, i_2, j_1, and j_2 satisfy $1 \le i_1 \le i_2 \le m$ and $1 \le j_1 \le j_2 \le n$. By $A(i_1{:}i_2$, $j_1{:}j_2)$ we mean the submatrix of A obtained by extracting rows i_1 through i_2

and columns j_1 through j_2. Thus, if

$$
A = \begin{bmatrix} 1 & 5 & 9 & 13 \\ 2 & 6 & 10 & 14 \\ 3 & 7 & 11 & 15 \\ 4 & 8 & 12 & 16 \end{bmatrix}
$$

then

$$
A(2{:}3, 3{:}4) = \begin{bmatrix} 10 & 14 \\ 11 & 15 \end{bmatrix}
$$

In many applications, it is necessary to pass designated submatrices as arguments to a subprogram. This can be done quite easily. For example, suppose we have the following subroutine for scalar-matrix multiplication:

```
      subroutine scale1( c, m, n, B, bdim )
c
      integer m, n, bdim
      real c, B(bdim,*)
c
c   Overwrites the m-by-n matrix B with cB.
c   The array B has row dimension bdim.
c
         do 10 j = 1,n
            do 5 i = 1,m
               B(i,j) = c*B(i,j)
    5       continue
   10    continue
         return
         end
```

Here is how $A(i_1{:}i_2, j_1{:}j_2)$ can be overwritten by $cA(i_1{:}i_2, j_1{:}j_2)$, assuming that adim houses the row dimension of A:

```
m = i2 - i1 + 1
n = j2 - j2 + 1
call scale1( c, m, n, A(i1,j1), adim )
```

To understand this, merely note that within scale1 we have the address mapping

$$addr[\,B(i,j)\,] \;=\; addr[\,B(1,1)\,] + bdim*(j-1) + (i-1)$$

$$= addr[\,A(i1,j1)\,] + adim*(j-1) + (i-1)$$

$$= addr[\;A(i1+i-1,j1+j-1)\;]$$

Thus, a reference to B(i,j) within scale1 is equivalent to a reference to A(i1+i-1,j1+j-1). As i ranges from 1 to m = i2-i1+1 and j ranges from 1 to n = j2-j1+1 we see that A(i1:i2,j1:j2) is scaled as required.

Stride

We now discuss another setting where "address reasoning" is important. Suppose we have the following subroutine for scalar-vector multiplication:

```
      subroutine scale2( n, c, v, incv )
c
      integer n, incv
      real c, v(*)
c
c  For j = 1,n, overwrites v(1+(j-1)incv) with
c              c*v(1+(j-1)incv)
c
      integer j, k
c
      k = 1
      do 10 j = 1,n
         v(k) = c*v(k)
         k = k + incv
   10 continue
      return
      end
```

The value of incv is referred to as the *stride*. The length of the scaling operation is passed through n. These two arguments, together with the "starting address" prescribed by the third argument, completely specify the vector to be scaled. Suppose c and v(1:8) are initialized. Then

```
call scale2( 4,c,v,2)      ⇒  v(1),v(3),v(5),v(7) scaled
call scale2( 4,c,v(2),2)   ⇒  v(2),v(4),v(6),v(8) scaled
call scale2( 2,c,v(3),4)   ⇒  v(3),v(7) scaled
```

A Note on Index Ranges

Variably dimensioned arrays need not be indexed from one in a subprogram. Moreover, a subprogram may index a given array differently than the calling program. To clarify these comments we recall that "index-from-one" is the default option. If

```
subroutine sub1( x, A, adim ... )
integer adim
real x(*), A(adim,*)
   :
```

then for addressing purposes, sub1 assumes that x(1) is the top of x and A(1,1) is the top of A. If within a calling program we have

```
   :
real y(0,63), B(-1:10,3:4)
   :
call sub1( y, B, 11,... )
   :
```

then inside sub1, x(i) is identified with y(i-1) and A(i,j) is identified with B(i-2,j+2).

Sometimes it is convenient to have unconventional subscripting within a subprogram. There are several ways to accomplish this but, given compiler variation, the most reliable method is to pass complete index range information for each "unconventionally" subscripted array:

```
subroutine sub2(x, p1, p2, A, q1, q2, r1, r2..)
integer p1, p2, q1, q2, r1, r2
real x(p1:p2), A(q1:q2,r1:r2)
    :
```

In this case, the location of element (i, j) of the matrix A is computed by

$$addr[\,A(i,j)\,] = addr[\,A(q1,r1)\,] + (j-r1)*(q2-q1+1)$$
$$+ (i-q1)$$

In a typical situation, we may have conventionally subscripted arrays in the main program,

```
real x(5 ), A(10,5)
```

and wish to regard them as indexed from zero in sub2. The proper way to invoke sub2 is then

```
call sub2( x, 0, 4, A, 0, 9, 0, 4,... )
```

Passing Arrays in Common Blocks

Common blocks provide another way to communicate to subprograms but, as we stressed in the previous section, their use should be minimized. However, when arrays are involved there is an additional reason to suppress their use and to pass arrays through argument lists. The major reason for this has to do with flexibility.

Common blocks can include array variables. For example, suppose in a main program we have

```
integer dim
parameter (dim = 10)
integer n
real A(dim,dim), v(dim), delta
common /matrix/A,v,delta,n,dim
```

This defines a common block with name matrix and it contains the array A, the vector v , and the scalars delta, n, and dim. This group of statements, or their equivalent, should appear in every subprogram that references these variables. As usual, it is not required that the names of the variables be the same in each subprogram, but they must be the same type and size and they must be listed in the same order. Arrays with variable dimensions *cannot* appear in common blocks. Note that the **parameter** statement is defining a constant, not a variable. Each subprogram that references a common block with arrays must explicitly declare those arrays, i.e.,

```
subroutine eg (...)
integer dim
parameter (dim = 10)
integer n
real A(dim,dim), v(dim), delta
common /matrix/A,v,delta,n,dim
```

It is tempting to delete the **parameter** statement and thereby let the main program "control" the size of the common arrays, but this is illegal. Each array in each common block must be dimensioned consistently in each subprogram in which it is used. This requirement discourages the use of common blocks in subroutines that are designed for flexible general library use. Returning to our example, changing the dimension of A in the main routine demands changing the corresponding **parameter** statements in each of the subprograms that use matrix.

We remark that there are circumstances in which it is not possible to pass an array through a subroutine argument list, thereby necessitating the placement of an array in **common**. For example, suppose that zero(f,a,b) is a library routine designed to find a zero of a continuous scalar function $f(x)$ on the interval $[a,b]$ given that $f(a)f(b) < 0$. Assume that the supplied function f has to be a function of a single real variable but that the function you are interested in has the form

$$f(x) = (ux - v)^{T}(ux - v) - 1$$

where u and v are vectors computed by the main program. To use zero it is necessary to house the data that defines f in a common block that is accessible to f:

```
program main
   :
real u(100), v(100)
integer n
common /fdata/u,v,n
   :

real function f(x)
real u(100), v(100)
integer n
common /fdata/u, v, n
real s
integer k
s = 0.
do 10 k = 1,n
    s = s + (u(k)*x - v(k))**2
10 continue
f = s - 1.
return
end
```

Problems for Section 1.6

1. Write a subroutine

$$saxpy(n, a, x, incx, y, incy)$$

that overwrites

$$x(1 + (i - 1)incx)$$

with

$$a\,x(1 + (i - 1)incx) + y(1 + (i - 1)incy)$$

for $i = 1{:}n$. Here, positive integers n, $incx$, and $incy$ are stored in integer variables n, incx, and incy and the appropriate portions of x and y are initialized.

2. Write a subroutine revers(n, x) that reverses the order of $x_1,\dots,x_n$.

3. We say that a matrix A has upper bandwidth p and lower bandwidth q if $a_{ij} = 0$, unless $j + q > i > j - p$. Write a subroutine

```
conv( A, adim, m, n, av, L )
```

that takes a conventionally stored A and stores the nonzero portion column by column in a singly subscripted array av.

4. Suppose a main program includes the statements

```
integer ix, iy, ia, ja
parameter( ix = 100, ia = 70, ja = 100, iy = 60)
real x(0:ix), y(0:iy), A(0:ia,0:ja)
```

Write a subroutine **matvec** that can compute the m-by-n matrix vector product $y = Ax$. Assume that m, n, x (0 : n - 1), and A (0 : m - 1, 0 : n - 1) are initialized. Subscript all arrays in **matvec** from zero.

1.7 INPUT AND OUTPUT

So far we have relied on the simple reading and writing of data. Now we examine the more sophisticated handling of I/O.

"read" and "write" Statements

The flow of input and output to and from a Fortran program requires interaction with the computer file system and a specification of the data's format. The formatted **read** and **write** statements are used for this purpose:

read({ *unit number* } , { *format number* }) { *list-of-variables* }

write({ *unit number* } , { *format number* }) { *list-of-variables* }

The first argument indicates where the data is coming from or going to, the second argument indicates the format of the data, and the variable list indicates the names of the variables involved. Thus, a single **write** statement could display on the user's terminal screen the contents of a double precision variable in some chosen style, say scientific notation with a four-digit mantissa.

Fortunately, and as we have been doing thus far, it is posssible to suppress many I/O details by using asterisks (*) for arguments in **read** and **write** statements. The asterisk invokes certain convenient default options and it is at this simple level that we begin our discussion.

List-directed "read"

A data file consists of records with any number of data items per record. Conceptually, it is handy to think of a record as a line. The items in a record may be separated by commas or blank spaces. Consider the two-record data file

```
10, 5, 6, 1.0
-1.0, 2.0
```

which we suppose to be accessible through some default convention. Assume i, j, and m are integer variables and that x, y, and z are real variables. If the following **read** statements are executed

```
read(*,*)i, j, m, x
read(*,*)y, z
```

then the integer variables i, j, and m are set to 10, 5, and 6 and the real variables x, y, and z are set to 1.0, -1.0, and 2.0 .

This example highlights a number of guidelines that pertain to the use of **read** statements. The type of input value must correspond to the type of variable to which it is being assigned. Data items 10, 5, 6 are integers (no decimal point) and correspond to integer variables i, j, m. Data items 1.0, -1.0, 2.0 are real, corresponding to real variables x, y, z. The format of the data is "directed" by the list of variables.

Each time a **read** statement is encountered a new record in the data file is accessed. So if there are more data items in a record than variables listed in the corresponding **read** statement, the extra data items are ignored. For example, if the above data file is changed to:

```
10, 5,  6,  1.0,  3, 5.0
-1.0,    2.0
```

then i, j, m, x, y, and z take on the same values as before but the data items "3" and "5.0" are ignored.

If the statements

```
read(*,*)i, j, m
read(*,*)x, y, z
```

are executed with data file

```
10,  5,  6,  1.0
-1.0,    2.0
```

then an error results even though there is enough data to go around. The reason for this is that the data item "1.0" is ignored and the program hangs up waiting for a value for z.

If the number of variables listed in a **read** statement exceeds the number of data items in the current input record, then subsequent records are read. For example, if the program segment

```
read(*,*)i, j, m, x
read(*,*)y, z
```

is executed with data file

```
10,   5
6
1.0
-1.0,    2.0
```

then all the variables are assigned. However, if the data file is replaced by

```
10, 5
6
1.0,   -1.0,    2.0
```

then y and z would not be assigned any new values. This is because upon reaching the statement

```
read(*,*)y, z
```

a new input record should be read but cannot be found .

List-directed "write"

The simplest form of this statement is

write (*,*) { *list-of-variables* }

Here, the contents of the variables listed are printed according to their type. For example, if i and j are integers and x and y are real then the statement

```
write(*,*)i, j, x, y
```

causes i and j to be printed with no decimal point and real numbers x and y to be printed either in positional notation (fixed form) or exponential notation, depending on the magnitude of the exponent.

In some cases it is useful to list the names of the variables along with their values. This can be done using single quotes:

```
write(*,*)'i =', i, 'j =', j, 'x =', x, 'y =', y
```

A feel for the appearance of list-directed output is best achieved through experimentation.

Formatted "read"

If the organization of the data in a file is known then the formatted **read** can be used. For example, suppose that every line of the data file consists of three integers, each occupying a *field* that is six places wide:

```
        3 8 2           1          5 6
|- - - - - -|- - - - - -|- - - - - -|

          1 1         1 0       1
|- - - - - -|- - - - - -|- - - - - -|

  1 2 3 4 5 6 - 1 2 3 4 5 6
|- - - - - -|- - - - - -|- - - - - -|
```

Of course only the digits are part of the data file. The markings are used to expose the spacing. Now consider the following formatted **read** statements, where all the variables in question have integer type:

```
        read(*,100)i,j,k
        read(*,100)m,n,o
        read(*,100)p,q,r
100     format( I6,I6,I6 )
```

The labeled **format** statement is interpreted as follows:

- There are three integer fields per line, each of the form "I6." The "I" means integer and the "6" means six spaces wide.

- Each integer occupies a six-column "field." This includes the sign bit. The + sign can be omitted. Thus, 123456 can be accommodated but -123456 cannot. The second number in the third record is -12345.

- All blanks are viewed as zeros so, for example, the third number in the second record is read as "100." Likewise, the third number in the third record is 600,000. Thus, each integer should be right-justified in its field unless the augmentation of low-order zeros is planned.

Floating point numbers can be read either in positional (fixed) form or exponential form. In the following example, the data file contains numbers in positional form:

```
   7 . 4              7 . 4
_  _ _ _ _ _ _|_ _ _ _ _ _

5 6 . 7 8 9 - 1 . 2 3 4 5
_ _ _ _ _ _|_ _ _ _ _ _
```

This data can be read into floating point variables as follows

```
      read(*,200)x,  y
      read(*,201)v,  w
200   format( F7.1,F6.1 )
201   format( F6.3,F7.4 )
```

Some comments on these particular formatted **read** statements:

- There are two F fields per line. These fields have widths 7 and 6 in the first line and widths 6 and 7 in the second. Note that decimal points and signs occupy a space.

- The "1" in "F7.1" means one place to the right of the decimal has been designated. In general, "Fi.j" means that j of the i places in the field are used for the fractional portion of the number.

- Unlike the I format case, the numbers do not have to be right-justified in their fields. For example, both x and y are assigned the same value above.

In general, formats of the form Fi.j should satisfy i ≥ j+2 because a space must be reserved for both the decimal point and the sign.

In the next example, the data file contains floating point numbers in exponential form:

```
 - 1 . 2 e - 1 2  5 6 7 . 1 2 3 e 1
```

This data can be read into real variables x and y as follows:

```
      read(*,100)x, y
  100 format(E9.1,  E10.3)
```

- There are two "E" fields having widths 9 and 10. The "i" in "Ei.j" indicates the field width and the "j" designates the length of the mantissa.

- The signs of the mantissa and the exponent and the decimal point occupy space. In an Ei.j format it should be the case that i ≥ j+5 so that there is enough room to handle the representation (assuming two-digit exponents).

- Trailing zeros are ignored in both the mantissa and exponent parts and so, for example, y is assigned the value 5671.23 .

Exponential format data can be read into double precision variables using the D format. The same rules that apply to the E format apply to the D format. Thus, if dx is a double precision variable then

```
           read(*,100) dx
    100    format(D15.8)
```

would successfully read in

$$321.12345678d+11$$

It is important to realize that the parameters in D, E, and F formats need not be related to the machine precision. If the number 1.2 is read into a double precision variable having a 56-bit mantissa, then the same decimal-to-binary conversion should take place whether a D10.1 or D25.15 format is used.

Formatted "write"

The formatted **write** allows for greater control of the appearance of output than the list-directed **write** statement. For example, if i and j are initialized integer variables and x and y are initialized real variables, then the program segment

```
           write(*,101)
           write(*,102)x, y, i, j
    101    format(1x,//,' Upon termination:')
    102    format(1x,///, 2x,   'x= ', F8.4, 2x, 'y= ',
         &  F.4, 2x, 'i= ', I5, 2X, ' j= ', I5)
```

produces output that looks something like the following:

```
Upon termination:

x= 12.3456   y= 12.3456 i= 12345 j= 12345
```

To appreciate why this is so, we summarize some facts about formatted **write**:

- The first character in an output line is for carriage-return. Here are the possibilities:

$$
\begin{array}{lll}
\text{blank} & \Rightarrow & \text{single spacing} \\
+ & \Rightarrow & \text{overwrite current line} \\
0 & \Rightarrow & \text{double spacing} \\
- & \Rightarrow & \text{triple spacing} \\
1 & \Rightarrow & \text{new page}
\end{array}
$$

- The slash (/) causes a blank line to be printed.

- Single quotes are used for names, titles, etc.

- 'iX' indicates that i spaces are to be skipped.

- 'Fi.j' indicates that fixed form, or positional notation, is to be used
  for the corresponding variable; the variable will occupy a total of
  i spaces, j of which occur after the decimal point. The
  parameters i and j should satisfy $i \geq j+2$.

The D and E formats can be used to output real and double precision values in
exponential notation. Thus, if D14.8 formats were used in the above example
instead of F8.4, then something like

```
Upon termination:

x= .123456D+02   y= .123456D+02   i= 12345 j= 12345
```

would result. Finally, we mention that expressions can be included within a
write statement, e.g.,

```
        write (*,200)  x*y
200     format ( 1x,  'xy product is ',   e13.6)
```

Location of "format" Statements

A **format** statement can appear anywhere in a program. If a given
format statement is used with only a single **read** or **write**, then it can be

placed adjacent to it. However, because **format** statements tend to be ugly and "disruptive" when reading a code, we advise collecting all the **format** statements in the program and placing them at the end just before the **end** statement.

Furthermore, we recommend distinctive labels for all the **format** statements that appear in a program. For example, if all the labels in the executable part of the code are between 1 and 999, then four-digit labels for the formats add a small amount of clarity and encourage the reader interested in the formats to look for their specification at the end of the program.

Some Shortcuts

Often a given format involves repetition, e.g.,

```
100    format( 2x, I2, 2x, I2, 2x, I2 )
```

This can be more conveniently expressed as

```
100    format( 3(2x,I2) )
```

Similarly,

```
100    format( I5,I5,/E10.2,/E10.2/ )
```

is equivalent to

```
100    format ( 2I5,2(/E10.2),/ )
```

The "data" Statement

Another way to input data is with the **data** statement. It can be used to assign values to variables and has the following form:

data {*list-of-variables*/{*list-of-values*}/,...

Here are some examples:

```
data n/100/, m/-5/, x/2.0/, y/2.0/, z/2.0/

data n,m/100,-5/, x,y,z/3*2.0/
```

These two data statements are equivalent in that they both result in the following assignments:

$$
\begin{aligned}
n &= 100 \\
m &= -5 \\
x &= 2.0 \\
y &= 2.0 \\
z &= 2.0
\end{aligned}
$$

Notice how identical values can be repeatedly assigned.

The **data** statement is performed exactly once -- just before execution of the program begins. Therefore, it is a convenient way to initialize variables. The **data** statement must be listed after all type declaration statements.

The **data** statement is handy for initializing vectors and matrices that are either very small or very regular in form. For example, the program segment

```
integer idim, jdim, kdim
parameter (idim = 4, jdim = 3, kdim = 2)
real A(idim,jdim),B(kdim,kdim), c(idim), d(kdim)
data A/12 * 1.0/, B(1,1)/1.0/
data B(1,2)/2.0/, c/4*0.0/, d(2)/1.0/
```

assigns a 4-by-3 matrix of ones to A, sets the first row of B to [1 2], assigns a 4-vector of zeros to c, and sets the second component of d to 1. Notice that the array names A and c appear in the **data** statement without arguments.

It is not possible to use the **data** statement to initialize a matrix which is a proper subset of a Fortran array, except on an element-by-element basis. In the above example this is done for the first row of the array B. The following is illegal and does *not* store the 2-by-2 zero matrix in A :

```
data A(1,1)/2*0.0/, A(1,2)/2*0.0/
```

It is important to realize that the **data** statement is performed just once before program execution begins. For this reason the **data** statement is usually not used within subroutines.

The **data** statement is quite different from the **parameter** statement. The latter is reserved for *constants*. In contrast, the variables listed in a **data** statement can be modified in subsequent **program** statements. We remark that early versions of Fortran did not support the **parameter** statement. Therefore the **data** statement was used for constants as well as for variable initialization. F77 programs should define constants using the **parameter** statement only.

The **data** statement cannot be used for variables within a common block However, there is a special kind of subroutine, called a "block data" subroutine, made for this purpose:

```
block data
integer dim
parameter (dim = 10)
integer n
real a(dim, dim), v(dim), delta
common /matrix/a,v,delta,n
data a/100*1.0/, v/10 * 2.0/, delta/1.0/, n/5/
end
```

The presence of this subroutine initializes the variables A, v, delta, and n, i.e., the variables that make up the common block named matrix.

Input and Output of Arrays

The **read** statement can be very convenient for inputting arrays. Suppose A is a two-dimensional array of size 50-by-40. The statement

```
read(*,*)A
```

attempts to read 2000 numbers and so the data file must contain at least 2000 numbers. The number of entries per record is arbitrary. This statement will cause A to be filled, column by column. Obviously this form of the **read** statement is not convenient when entering data for a matrix A that is strictly embedded in the array A since it demands enough data to fill up the entire array. To handle this eventuality we can use

```
read(*,*)((A(i,j), i=1,m), j=1, n)
```

where we assume that the row and column dimensions of the matrix *A* are stored in m and n. Here the array A is initialized column by column. Of course if the data is naturally generated in a row-by-row fashion then

```
read(*,*)((A(i,j), j=1,n), i=1,m)
```

also does the job. However, bear in mind that this order may result in poor performance as it does not access contiguous blocks of memory. Try to arrange input in column-oriented fashion. See §1.9.

Of course the default options, (*,*), need not be used. The above constructions can be used with **format**. For example,

```
   read(*,10)((A(i,j), i=1,m), j=1,n)
10 format(10I5)
```

indicates that the matrix *A* is read according to format 10. That is, each line of data consists of ten data entries, each of which is an integer occupying five columns.

The output of an entire array A(idim, jdim) can be achieved as follows:

```
write(*,*)A
```

or, using formatted **write**,

```
   write(*,10) A
10 format( 6 F7.4)
```

which results in six numbers per line. It is important to realize that when A is referred to without arguments then the entire *array* is printed, not just the contained matrix or vector. Row *i* of an *m*-by-*n* matrix *A* can be printed by the statement

```
write(*,*)(A(i,j),j=1,n)
```

One must remember that output lines have a finite width, e.g., 80 characters. If a program attempts to print a line that is wider than the maximum, then the output will "spill over" into one or more succeeding records. This often happens when attempting to print matrices that have a fair number of columns. Care must be exercised to produce aesthetic output.

Problems for Section 1.7

1. Write a subroutine `month(length,start)` that reads the number of days in a month into `length` and the "starting day" of the month (1=Sunday,...,7 = Saturday) into `start` and prints a calendar. For example, for input (30,5) the following would be printed:

```
              1   2   3
   4   5   6   7   8   9  10
  11  12  13  14  15  16  17
  18  19  20  21  22  23  24
  25  26  27  28  29  30
```

2. Write a subroutine `print( A, adim, i, j )` that prints the submatrix $A(i:i+2, j:j+2)$. Assume that A is an integer matrix with entries in the interval [-999,999] and that i and j are no larger than 99. Sample output (for input $i = 7, j = 12$):

```
                        123    567   -892
      A(7:9,12:14)   =  187    728    823
                        632     76     -9
```

3. Assume that v is an initialized real n-vector stored in the array v. Write a subroutine `nicepr(v)` that computes an integer e (if possible) so that $10^{e-3} < | v(i) | < 10^{e+3}$ and then prints v as follows:

```
v = 10**e *
              xxx.xxx
                :
              xxx.xxx
```

where the components are in F7.3 format. If no such e exists then the components of v should be printed in a column using E format.

1.8 COMPLEX ARITHMETIC

Just about everything we have presented carries over to the case of complex variables. We summarize the key constructions.

Declaring Complex Variables

Just as the complex number $z = x + iy$ $(i^2 = -1)$ is an ordered pair (x,y) of real numbers, so are complex variables an ordered pair of floating point variables. The declaration

```
complex z
```

sets aside two real variables for representing a complex number: one for the real part and one for the imaginary part. Physically, it is helpful to picture z as a pair of adjacent real variables. Complex arrays are also possible. For example,

```
integer adim
parameter(adim = 10)
complex A(adim,adim)
```

sets up a 10-by-10 complex array called A . Because complex variables take up twice as much space as real variables, memory constraints sometimes pose a problem when large complex arrays are involved.

Useful Built-in Functions

In a typical situation one often has to "set up" a complex number from two real numbers. This can be accomplished with **complx**. Assuming that x and y are stored in real variables x and y, the statement

```
z = complx(x,y)
```

assigns $x + iy$ to z . Likewise,

```
x = real(z)
y = aimag(z)
```

extract the real and imaginary parts of the complex number stored in z. Complex conjugates can be obtained as follows:

```
w = conjg(z)
```

while the functions **cabs, cexp, clog, ccos**, and **csin**, respectively, compute the exponential, logarithm, cosine, and sine of a complex argument. Note that **cabs** is a real-valued function.

Using Complex Variables

Complex variables can be manipulated just as can real and double precision variables. If p and z are complex then the assignment

```
p = z**2 + z
```

is equivalent to

```
x  =  real(z)
y  =  aimag(z)
u  =  x*x - y*y + x
v  =  2.0e0*x*y + y
p  =  complx(u,v)
```

The function **complx** should be used to specify complex constants. Thus, if $p(z) = z^2 + z + c$, where $c = 3 + 4i$, then

```
p = z*z + z + complx(3.0e0,4.0e0)
```

Input/Output

The input and output of complex data is fairly straightforward. To begin with,

```
write(*,*)z
```

is equivalent to

```
write(*,*)real(z), aimag(z).
```

Regarding formatted I/O, the D, E, and F formats can be used with the understanding that a complex variable is really two floating point variables. Thus, if z is complex

```
         write(*,100)z
100      format('real(z)=',f10.7,'imag(z)=' f10.7)
```

is equivalent to

```
         x = real(z)
         y = aimag(z)
         write(*,100)x,y
   100   format('real(z)= ',f10.7,'imag(z)=' f10.7)
```

where x and y are real.

Avoiding Complex Arithmetic

It is sometimes possible to avoid complex data types even though complex arithmetic arises. The way to do this is to "simulate" complex arithmetic using real arithmetic. Consider the following function that computes the absolute value of the largest root of the real quadratic equation $x^2 + 2bx + c = 0$.

```
         real function maxrt( b, c )
         real b, c
c
c  Computes max{ abs(r1),abs(r2) } where
c  r1 and r2 are roots of the real quadratic
c  x2 + 2bx + c = 0.
```

```
c
      d = b*b - c
      if (d .GE. 0.0e0)   then
c
c     Real  roots
c
         maxrt = abs ( -b + sign(sqrt(d),-b))
c
      else
c
c     Complex  roots
c
         maxrt = sqrt ( b*b - d)
c
      endif
      return
      end
```

This is preferable to the code

```
      d =    csqrt ( complx ( b*b + cc , 0.0e0))
      maxrt = max (cabs (-b+d) , cabs (-b-d))
```

Another situation concerns products of the form real times complex. Suppose that we want to scale a complex vector x by a real scalar a. The code

```
      do 10 k = 1,n
         x(k)  = complex (a, 0.0d0) *x(k)
10 continue
```

requires twice as much arithmetic as

```
      do 10 k =1,n
         x(k)=complex (a*real(x(k)),  a*aimag(x(k)))
10 continue
```

Lastly, we mention that complex data often comes in "conjugate pairs" and so allows for compact representation. For example, a complex Hermitian matrix $A = (a_{ij})$ has the property that a_{ij} is the conjugate of a_{ji}. Thus, A could be stored in a real array A with the convention that A(i,j) contains the real part of a_{ij} if $i \leq j$ and the imaginary part of a_{ij} if $i > j$.

Problems for Section 1.8

1. Write a subroutine

```
FFTMAT ( A, n, adim )
```

that sets A(p,q) to $w^{(p-1)(q-1)}$, where $w = cos(2\pi/n) + i \, sin(2\pi/n)$ and p and q satisfy $1 \leq p,q \leq n$. Exploit the fact that w is an nth root of unity and so $w^m = w^r$, where $r = m \bmod n$. Assume that A is a complex array with row dimension stored in adim.

2. Write a subroutine

```
POLAR ( z, r, theta )
```

that converts a complex number of the form $z = x + iy$ to polar form $z = r \, e^{i\theta}$.

3. Write a subroutine

```
HOUSE ( n, v)
```

that overwrites the complex n-vector v with a unit 2-norm complex vector u that is in the direction $v + e^{-i\theta} \| v \|_2 \, e_1$, where e_1 is the first column of the identity and $v_1 = | v_1 | e^{i\theta}$.

1.9 PROGRAMMING TIPS

Throughout the previous sections we have given guidelines for writing "good code." We continue in this vein by singling out some additional techniques that help produce readable code that is both efficient and reliable. It should be stressed that the art of writing good F77 programs is in large measure, just that--an art. We cannot completely delineate what it takes to write good code. But by scratching the surface here with a few examples we can highlight some of the major issues.

Intermediate Variables for Clarity and Efficiency

Sometimes the evaluation of complicated expression is made more efficient and understandable if intermediate variables are employed and care is taken in the arrangement of the computations. For example, if $P = (r, a_1, b_1)$ and $Q = (r, a_2, b_2)$ are the spherical coordinates of two points, then the distance d between them is specified by

$$d^2 = [\, r \sin(b_1)\cos(a_1) - r\sin(b_2)\cos(a_2)\,]^2 +$$
$$[\, r \sin(b_1)\sin(a_1) - r\sin(b_2)\sin(a_2)\,]^2 +$$
$$[\, r \cos(b_1) - r\cos(b_2)\,]^2 .$$

Here are three ways to compute d. The first is a straight encoding of the formula, the second exploits some common subexpressions and identifies the (x,y,z) distances, and the third exploits some trigonometric identities:

Method A

```
d   =   sqrt (
&  (  r*sin(b1)*cos(a1)  -  r*sin(b2)*cos(a2)  )**2 +
&  (  r*sin(b1)*sin(a1)  -  r*sin(b2)*sin(a2)  )**2 +
&  (  r*cos(b1)  -  r*cos(b2)  )**2 )
```

Method B

```
s1 = sin(b1)
```

87

```
s2 = sin(b2)
xdist = s1*cos(a1)   -   s2*cos(a2)
ydist = s1*sin(a1)   -   s2*sin(a2)
zdist = cos(b1) - cos(b2)
d = r*sqrt( xdist**2  + ydist**2 + zdist**2 )
```

Method C

```
t = cos(a1 - a2)
d = r*sqrt(1. - t*cos(b1-b2) -
&                 cos(b1)*cos(b2)*(1. - t))
```

One criterion for efficiency in this calculation is the number of trigonometric evaluations. Methods A, B, and C require 10, 8, and 4 trigonometric evaluations, respectively. This distinction is particularly important if the computations appear inside a loop. On the other hand, Method C is quite far removed from the "language" of spherical coordinates and so is somewhat more difficult to read than methods A and B. This deficiency could be addressed with sufficient comments. But the reader should be aware that in many other situations readability and efficiency may be difficult to reconcile. A good example of this is the packed storage scheme for symmetric matrices that we discussed in §1.5. It is efficient with respect to memory, but the nonstandard subscripting makes packed storage programs difficult to follow.

Sometimes a temporary array can lead to a more efficient computation. For example, to set up an n-by-n matrix F with $f_{ij} = \exp(-i - .5j)$ we can reduce the number of exp evaluations by an order of magnitude by introducing a pair of linear work arrays:

```
      do 10 j = 1,n
         x(j)  = exp(-float(j))
         y(j)  = exp(-.5 *float(j))
   10 continue
      do 20 j = 1,n
         do 15 i = 1,n
            F(i,j)  = x(i)*y(j)
   15    continue
   20 continue
```

Because of memory concerns, a work array should "earn its keep." But the matter should be kept in perspective. For example, a few linear work arrays hardly matter in a program that has two-dimensional arrays of the same order.

Checking Input Parameters

It is a good idea to check input arguments to a subprogram to make sure that they fall within the range of "permissible" values. Consider a subroutine

```
mmult ( A, adim, ma, na, B, bdim, mb, nb, C, cdim )
```

designed to compute the product $C = AB$, where A is an m_a-by-n_a matrix and B is an m_b-by-n_b matrix. A good way to begin the code is with the following checks and messages:

```
if ( na .NE. mb ) then
    write(*,*) 'Product AB not defined'
    return
elseif (cdim .LT. m)
    write(*,*) 'Array C not big enough.'
    return
endif
```

Test Most Likely Conditions First

In an **if-then-elseif** construction, arrange the conditions in decreasing order of likelihood to minimize conditional testing at execution time. To illustrate, consider the program segment

```
    do 10 j = 1,n
        do 5 i = 1,n
            if ( i .LT. j ) then
                A(i,j) = -1.
            elseif (i .GT. j ) then
                A(i,j) = 0.
            elseif (i .EQ. j ) then
                A(i,j) = 1.
            endif
 5       continue
10  continue
```

This sets up an n-by-n upper triangular matrix that has 1's on the diagonal and -1's above the diagonal. It is easy to show that $n(3n+1)/2$ conditionals are tested during execution. If the "i .EQ. j" conditional were to come first, almost twice as many comparisons would be required.

The example has further interest because it can be written in such a way as to avoid all "i-j" comparisons:

```
      do 10 j = 2,n
         do 5 i = 1,j-1
            A(i,j) = -1
  5      continue
 10 continue
      do 20 j = 1,n
         A(j,j) = 1
 20 continue
      do 30 j = 1,n-1
         do 25 i = j+1,n
            A(i,j) = 0
 25      continue
 30 continue
```

The price paid for this rearrangement is a slight increase in the number of loops. However, for large n this is not important.

Integer Arithmetic

It is important to be aware of the integer arithmetic overheads associated with subscripting and loop execution. Integer arithmetic is *not* trivial compared to floating point arithmetic. Because more attention is typically paid to the latter, it often happens that a program's performance is degraded because of careless looping and subscripting. Suppose that x(0:n-1) and y(0:n-1) are initialized and that we want to compute

$$s_k = x_0 y_k + \dots + x_{n-k-1} y_{n-1} + x_{n-k} y_0 + \dots + x_{n-1} y_{k-1}$$

for $k = 0,\dots,n-1$. In practice one would do this using fast Fourier transforms. However, if we approach the problem with a straightforward double-loop

encoding we can then illustrate a couple of widely used "tricks." We begin with
the naive algorithm:

```
      do 30 k = 0,n-1
         s(k) = 0.
         do 10 i = 0,n-k+1
            s(k) = s(k) + x(i)*y(i+k)
10       continue
         do 20 i = n-k,n-1
            s(k) = s(k) + x(i)*y(i-n+k)
20       continue
30 continue
```

In assessing the efficiency of the computation we focus on the "s(k)" updates,
as they are the most deeply nested statements. Two maneuvers can significantly
reduce the amount of subscripting , i.e., the amount of integer arithmetic. First,
we can build the running sum in a scalar variable. This avoids having to index
into the s vector until the accumulation of s_k is complete. The other economy
results by getting the "k-n" computation out of the 20-loop. (This assumes a
"stupid" compiler.) With these alterations we obtain

```
      do 30 k = 0,n-1
         t = 0.
         do 10 i = 0,n-k+1
            t = t + x(i)*y(i+k)
10       continue
         kmn = k - n
         do 20 i = n-k,n-1
            t = t + x(i)*y(i+kmn)
20       continue
         s(k) = t
30 continue
```

Generality versus Efficiency

Program generality can threaten efficiency unless care is exercised. Consider the following function that computes the dot product of two *n*-vectors:

```fortran
      real function dot1(n,x,y)
      integer n
      real x(*),  y(*)
      integer i
      real s
      s = 0.
      do 10 i = 1,n
         s = s + x(i)*y(i)
   10 continue
      dot1 = s
      return
      end
```

Note that the components of the vector arguments must be "adjacent" in memory. The function could not be used to compute

$$s = x(1)*y(1) + x(3)*y(3) + x(5)*y(5)$$

because the *stride* is two. (The stride of a vector refers to the spacing between the components.) To rectify this we expand the argument list so that stride information can be passed:

```fortran
      real function dot2( n,  x,  incx,  y,  incy )
      integer n,  incx,  incy
      real x(*),  y(*)
      integer i,  ix,  iy
      real s
      ix = 1
      iy = 1
      s = 0.
      do 10 i = 1,n
         s = s + x(ix)*y(iy)
```

```
      ix = ix + incx
      iy = iy + incy
10 continue
   dot2 = s
   return
   end
```

A call of the form

```
s = dot2(3,x,2,y,2)
```

is thus equivalent to

```
s = x(1)*y(1) + x(3)*y(3) + x(5)*y(5)
```

Note that dot2 is a more general dot product routine than dot1, but that it involves considerably more integer arithmetic. For unit stride dot products, we are better off with dot1. If unit stride dot products are sufficiently common (and they usually are) then dot2 is not that attractive. However, the efficiency of dot1 for the unit stride case and the generality of dot2 can be efficiently combined as follows:

```
      real function dot3( n, x, incx, y, incy )
      integer n, incx, incy
      real x(*), y(*)
      integer i, ix, iy
      real s
      s = 0.
      if ( incx .EQ. 1 .AND. incy .EQ. 1) then
         do 5 i = 1,n
            s = s + x(i)*y(i)
   5     continue
      else
         ix = 1
         iy = 1
         do 10 i = 1,n
            s = s + x(ix)*y(iy)
            ix = ix + incx
            iy = iy + incy
  10     continue
      endif
```

```
dot3 = s
return
end
```

The program is longer but it is still clear. We mention that this discussion is based on the design of the BLA subprogram SDOT discussed in the next chapter.

Machine-independent Testing For Roundoff Noise

The unit roundoff u in a floating point system that has t-bit mantissas is defined by

$$u = 2^{1-t}.$$

This is (roughly) the smallest positive floating point number for which $1 + u > 1$ in floating point arithmetic.

If x and y are nonnegative floating point numbers then $x < uy$ implies that x is no bigger than the least significant bit in y's floating point representation. On a machine with 56-bit floating point arithmetic, a program that tests for this situation might be structured as follows:

```
    :
parameter ( u = 2**(-55) )
    :
if ( x .LT. u*y ) then
    :
else
    :
endif
    :
```

The trouble with this is that the program only works in 56-bit floating point systems. If we use the program on a different machine then the parameter u must be changed accordingly. This is inconvenient and prone to errors.

A better solution is based on the fact that if $x + y = y$ in floating point, then x must be very small compared to y. In fact, very reasonable assumptions about the underlying arithmetic allow us to conclude that if $x + y = y$ in floating point then $x < uy$. We therefore obtain a much more attractive *machine-independent* version of the above **if-then-else** construction:

```
z = x + y
if ( z .GT. y ) then
    :
else
    :
endif
```

The reason for the intermediate variable z is to force the storage of the sum $x + y$. Otherwise the sum and test might take place in extended precision registers and give a false impression about the size of x. (This ploy works with most compilers.)

Termination Criteria

The termination of an iterative process requires care. Here is a real function that sums the Taylor series for $\exp(x)$ until the kth term in absolute value is less than a tolerance tol times the kth partial sum:

```
real function exp1(x,tol)
real x,tol
real s, term
integer k
s = 1.
term = x
do 10 k = 1,30
    s = s + term
    if ( abs(term) .LT. tol*abs(s) )
        exp1 = s
        return
    endif
    term = term*(x/real(k))
10 continue
exp1 = s
return
end
```

Notice that the function is unwilling to go beyond the thirtieth partial sum.

There are two things wrong with `exp1`. In situations like this with "indefinite termination" it is better to use the **while** construct. Second, the user-defined tolerance may be too stringent given the underlying machine precision. This implies that more iterations may be performed than are necessary. Here is a revision that addresses these two problems:

```
real function exp2(x,tol)
real x,tol
integer kmax, k
parameter (kmax = 30)
real s, snew
logical big
s = 1.
k = 0
term = x
snew = s + term
big = term .GE. tol*s  .AND. snew .NE. s
10 if ( k .LE. kmax .AND. big )  then
      s = snew
      k = k+1
      term = term*(x/float(k))
      snew = s + term
      big = term .GE. tol*s  .AND. snew .NE. s
      goto 10
   endif
   exp2 = s
   return
   end
```

In general, it is a good idea to process user-defined tolerances in this way because programmers typically have an unrealistic view of floating point accuracy.

Computing Small Corrections

Suppose a and b are nearby floating point numbers. It is better to compute the midpoint of the interval [a,b] as a small correction to a,

$$m = a - (b - a)/2$$

rather than with the formula

$$m = (a + b)/2 \ .$$

This example is discussed in Kahaner, Moler, and Nash [5, chap. 7] and has to do with the peculiarities of floating point arithmetic.

Pay Attention to the Innermost Loops

Suppose that we want to evaluate the polynomial

$$p(x) = a_0 + a_1 x^2 + a_2 x^4 + \ldots + a_n x^{2n}$$

at some point $x = z$. Using Horner's rule we obtain:

```
    s = a(n)
    do 10 k = n-1:-1:0
        s = s*z*z + a(k)
10 continue
```

Recognize, however, that we can halve the number of floating point multiplies by getting the "z*z" computation outside the loop:

```
    s = a(n)
    w = z*z
    do 10 k = n-1:-1:0
        s = s*w + a(k)
10 continue
```

As a general rule loop bodies should always be searched for redundancies. Of course, a smart compiler can sometimes detect and remove computational redundancies.

Guarding against Overflow

Consider the computation of a cosine-sine pair (c,s) such that $-sx + cy = 0$, where x and y are given real numbers. Here is a naive solution:

```
d = sqrt( x**2 + y**2)
if d .GT. 0
    c = x/d
    s = y/d
else
    c = 1.
    s = 0.
endif
```

A problem with this is that overflow might occur in the computation of d because the square of x or y might exceed the largest floating point number. A way around this difficulty is to work with scaled versions of x and y:

```
m = abs(x)  + abs(y)
if m   .NE.  0   then
    x = x/m
    y = y/m
    d = sqrt( x*x + y*y )
    c = x/d
    s = y/d
else
    c = 1.
    s = 0.
endif
```

The procedure now involves four divides, two multiplies and two adds, and one compare. Some of these additional operations can be avoided by exploiting some connections between $c, s, tan(\theta) = y/x$ and $ctn(\theta) = x/y$:

```
if abs(y)  .GT.  abs(x)
    t = x/y
    s = 1./sqrt(1 + t*t)
    c = s*t
```

```
elseif (abs(x) .GT. 0.)
   t = y/x
   c = 1./sqrt(1. + t*t)
   s = t*c
else
   c = 1.
   s = 0.
endif
```

Column-oriented versus Row-oriented Algorithms

There are times when it is absolutely crucial for the Fortran programmer to be aware of the manner in which space is allocated for two-dimensional arrays. Consider the matrix-vector multiplication $y = Ax$, where A is an m-by-n matrix and x is an n-vector. It is better to compute y using

```
      do 10 i   =   1,m
         y(i) = 0.
10 continue
      do 30 j=1, n
         do 20 i = 1, m
            y(i) = y(i) + A(i,j)*x(j)
20       continue
30 continue
```

than by using the mathematically equivalent program segment

```
      do 30 i = 1,m
         y(i) = 0.0
         do 20 j = 1,n
            y(i) = y(i) + A(i,j)*x(j)
20       continue
30 continue
```

The reason has to do with the way arrays are allocated. Because arrays are stored by column and because the inner loop in the first example varies the row index, the first code results in contiguous cells of storage being accessed. This is a great advantage on many machines since localized memory references can mean less page swapping to and from fast memory, and thus more efficient code. Moreover, accessing contiguous memory cells is an important concern on vector/pipeline computers.

Whenever possible have innermost loops vary the row index of a referenced two-dimensional array.

Problems for Section 1.9

1. Consider a pyramid with a square base and height h. Assume that the base has vertices $A, B, C,$ and D whose spherical coordinates (r, θ, ϕ) are given by

$$
\begin{array}{rcl}
A & = & (\ 1\ ,\ \theta\ ,\ 0\ ) \\
B & = & (\ 1\ ,\ \theta\ ,\ \pi/2\ ) \\
C & = & (\ 1\ ,\ \theta\ ,\ \pi\ ) \\
D & = & (\ 1\ ,\ \theta\ ,\ 3\pi/2\ )
\end{array}
$$

where $-\pi/2 \le \theta \le \pi/2$. Assume that the apex is at $(1,\pi/2,0)$. Write a single precision program that reads in h and θ and prints the pyramid's volume V and surface area S. Arrange your program so that it involves a minimum number of sine/cosine evaluations.

2. Suppose A (n-by-n) and x (n-by-1) are given. Give an efficient algorithm for computing

$$
s\ =\ x^T A\,x\ =\ \sum_{i=1}^{n} \sum_{j=1}^{n} a_{ij}\, x_i\, x_j
$$

3. How could the matrix multiply subroutine mmult be used to compute $M = ABC$, where A (m-by-n), B (n-by-p), and C (p-by-r) are rectangular matrices whose row and column dimensions are known.

Appendix. FORTRAN 77 BUILT-IN FUNCTIONS

Fortran 77 contains a number of built-in, or intrinsic, functions. Such a subprogram is invoked like any other function subprogram (except the user does not have to supply it). The intrinsic functions are divided into two classes: specific and generic. A specific function requires arguments of a particular type and returns a predefined type. A generic function accepts arguments of any appropriate type; the return value is determined by the types of the arguments. Just about all of the functions below are generic.

A large number of specific functions exist but are not listed because the generic feature makes their use obsolete. For example, if x and root are **double precision** variables then

```
root = dsqrt(dabs(x))
```

computes the double precision square root of *x*. This is because **dabs** is the intrinsic function for determining the absolute value of a double precision variable, and **dsqrt** is the intrinsic function for a double precision square root The following statement produces the same result:

```
root = sqrt(abs(x))
```

Integer/Real/Double Precision Functions

These functions accept either integer, real, or double precision arguments. They return a value of the same type except **nint**, which always returns an integer.

int(x)	sign of x times largest integer $\leq$ abs(x)
nint(x)	round to nearest integer
mod(x,y)	x - y*int(x/y)
sign(x,y)	sign of y times abs(x)
max(x1,x2,...)	max{x1,x2,...}
min{x1,x2,...)	min{x1,x2,...}

Real/Double Precision Functions

These functions accept either real or double precision value and return a value of the same type.

abs(x)	absolute value
sqrt(x)	square root
exp(x)	exponential
log(x)	natural logarithm
log10(x)	base ten logarithm
cosh(x)	hyperbolic cosine
sinh(x)	hyperbolic sine
tanh(x)	hyperbolic tangent
cos(x)	cosine
sin(x)	sine
tan(x)	tangent
acos(x)	arccosine
asin(x)	arcsine
atan(x)	arctangent

Complex

These functions accept complex arguments. **cabs, real,** and **aimag** return real values. The rest return complex values.

cabs(z)	absolute value
real(z)	real part
aimag(z)	imaginary part
conj(z)	conjugate
cexp(z)	exponential
clog(z)	natural logarithm
ccos(z)	cosine
csin(z)	sine

String

Length, index, and **ichar** accept character arguments and return integers. **Char** applies to integer types and returns a character.

length(s)	length of string
index(s1,s2)	location of s2 in s1
ichar(s)	convert character to integer
char(i)	convert integer to character

Numerical Type Conversion

These functions accept integer, real, and double precision types and return values of the designated type:

int(x)	convert to integer
real(x)	convert to single precision
dble(x)	convert to double precision
cmplx(x,y)	convert to complex

Chapter 2

The BLAS

In this and the next chapter we discuss LINPACK, a package for solving linear system and least squares problems. The "high-level" routines in LINPACK (to be discussed in the next chapter) all access a collection of "low-level" routines that perform elementary operations on vectors. These low-level routines are known as the Basic Linear Algebra Subprograms, or BLAS. There are BLAS for bookkeeping, vector operations, norm calculations, and Givens rotations. In general, each BLA subprogram is implemented in real single precision, real double precision, complex single precision, and complex double precision. For simplicity, we describe the real single precision BLAS in §§2.1-2.4 . In §2.5 the complex and double precision implementations are discussed. The BLAS are discussed in C.L. Lawson, R.J. Hanson, F.T. Krogh, and O.R. Kincaid [6] and in the LINPACK manual of Dongarra, Bunch, Moler, and Stewart [1].

We mention that the BLAS covered in this chapter are what are called "Level-1" BLAS. They involve vector operations: $O(n)$ input and $O(n)$ work. Level-2 and Level-3 BLAS also exist and are increasingly important software building blocks in various high-performance computing environments. See Golub and Van Loan [4], J. Dongarra, J. Du Croz, S. Hammarling, and R.J. Hanson [2], and J. Dongarra, J. Du Croz, I.S. Duff, and S. Hammarling [3] for more details. Level-2 BLAS involve matrix-vector operations such as matrix-vector multiplication and rank-1 updates: $O(n^2)$ data and $O(n^2)$ work. Matrix-matrix multiplication is one of the Level-3 BLAS : $O(n^2)$ data and $O(n^3)$ work.

2.1 BOOKKEEPING OPERATIONS

Two of the BLAS deal with the movement of linear algebraic data. These are the subroutines SCOPY and SSWAP. To overwrite a vector y with a vector x, invoke SCOPY:

```
call SCOPY( n, x, incx, y, incy )
```

To interchange their contents, use SSWAP :

```
call SSWAP( n, x, incx, y, incy )
```

The effective dimension of x and y is passed through n and their "stride" or increment through incx and incy, respectively. The x and y arguments specify the "tops" of the vectors x and y. We clarify this through examples. If x(1:100) and y(1:50) are initialized then

call SCOPY(20,x,1,y,1)	$\Rightarrow$ $x(i) \leftarrow y(i)$	$i = 1{:}20$
call SCOPY(15,x(10),1,y(20),1)	$\Rightarrow$ $y(19+i) \leftarrow x(9+i)$	$i = 1{:}15$
call SCOPY(10,x(11),1,x(1),1)	$\Rightarrow$ $x(i) \leftarrow x(10+i)$	$i = 1{:}10$
call SCOPY(50,x(1),2,y(1),1)	$\Rightarrow$ $y(i) \leftarrow x(-1+2i)$	$i = 1{:}50$
call SSWAP(20,x,1,y,1)	$\Rightarrow$ $x(i) \leftrightarrow y(i)$	$i = 1{:}20$
call SSWAP(10,x(20),3,y(2),2)	$\Rightarrow$ $x(17+3i) \leftrightarrow y(2i)$	$i = 1{:}10$

Zero increments are perfectly legal and occasionally useful. For example, a handy way to set up an n-dimensional zero vector is by the call

```
call SCOPY(n, 0, 0 , y, 1 )
```

Several other more detailed examples follow that illustrate SSWAP and SCOPY.

Example 2.1-1 (Matrix Copy)

```
        subroutine MATCOP ( m, n, B, bdim, A, adim )
        integer m, n, bdim, adim
        real B(bdim,*), A(adim,*)
c
c   Overwrites the m-by-n matrix A with the
c   m-by-n matrix B.
c
c   On entry:
c       B        real(bdim,n)   The matrix B.
c       bdim     integer        Row dimension of the array B.
c       A        real(adim,n)   The matrix A.
c       adim     integer        Row dimension of array A
c       m        integer        Row dimension of matrices B and A
c                               Must have m <= adim, bdim
c       n        integer        Column  dimension of matrices
c                               B and A.
c
c   On exit:
c       A        Returns a copy of the matrix B.
c
c   Subprograms called: SCOPY
c
c   Local variables:
        integer k
c
c   Copy B to A, column by column.
c
        do 10 k = 1,n
            call SCOPY( m, B(1,k), 1, A(1,k), 1 )
    10   continue
c
        return
        end
```

Example 2.1-2 (*Matrix Transpose*)

```
      subroutine TRANSP( n, A, adim )
      integer n, adim
      real A(adim,*)
c
c  This subroutine overwrites an n-by-n matrix A with its
c  transpose.
c
c  On entry:
c
c     A        real(adim,n)   The matrix A.
c     adim     integer        Row dimension of the array A.
c     n        integer        Order of the matrix A.
c
c  On exit:
c     A                       Returns the matrix A^T.
c
c  Subprograms called: SSWAP
c
c  Local variables
      integer k
c
c  Swap the kth upper triangular row with the
c  kth lower triangular column.
c
      do 10 k = 1, n-1
         call SSWAP( n-k, A(k+1,k), 1, A(k,k+1), adim )
   10 continue
c
      return
      end
```

Example 2.1-3 (Circulant Matrix)

```
      subroutine CIRCUL( n, C, cdim, v )
      integer n, cdim
      real C(cdim,*), v(*)
c
c  This subroutine sets up an n-by-n circulant matrix C
c  whose kth column is given by
c
c      [ v(k) v(k+1) ... v(n) v(1) ... v(k-1) ]^T
c
c  On entry:
c      v        real(n)        The n-vector v.
c      n        integer        Order of the matrix C.
c      cdim     integer        Row dimension of the array C.
c                              Must have n <= cdim
c
c  On exit:
c      C        real(cdim,n)   Returns the circulant matrix C.
c
c  Subprograms called:  SCOPY
c
c  Local variables
      integer k
c
c  Set up the circulant matrix C, column by column.
c
      call SCOPY( n, v(1), 1, C(1,1), 1 )
      do 10 k = 2,n
          call SCOPY( n-1, C(2,k-1), 1, C(1,k), 1 )
          C(n,k) = C(1,k-1)
   10 continue
c
      return
      end
```

Problems for Section 2.1

1. Would a single call to SCOPY of the form

```
call SCOPY( m*n, A(1,1), 1, B(1,1), 1 )
```

work in Example 2.1-1 ?

2. Write a Fortran subroutine

```
SYMCOP( n, A, adim, B, bdim )
```

that accepts as input an n-by-n symmetric matrix A and copies the upper triangular part into an array B.

3. Write a Fortran subroutine

```
PERM( n, A, adim, work )
```

that overwrites a given n-by-n matrix A with the matrix AP, where P is the permutation $P = [\, e_2, ..., e_n, e_1 \,]$ and $I_n = [\, e_1, ..., e_n \,]$.

4. Write a Fortran subroutine

```
HAM( n, A, adim, F, fdim, G, gdim, H, hdim )
```

that computes the $2n$-by-$2n$ *Hamiltonian* matrix

$$ H = \begin{bmatrix} A^T & F \\ G & -A \end{bmatrix} $$

given $A \in R^{n \times n}$, $F \in R^{n \times n}$ (symmetric), and $G \in R^{n \times n}$ (symmetric).

2.2 VECTOR OPERATIONS

The BLAS considered in this section perform the following elementary operations:

1. Multiplying a vector by a scalar (SSCAL)
2. Adding a multiple of one vector to another (SAXPY)
3. Computing the inner product of two vectors (SDOT)

After we define these three subprograms, several examples are given to clarify their use.

Scaling Vectors

The subroutine SSCAL is used to form products of the form ax, where a is a scalar and x is a vector:

```
call SSCAL( n, a, x, incx )
```

The "top" of the vector to be scaled is passed through x, the scale factor is a, the effective dimension of the scaled vector is n, and incx is the increment. To illustrate how these parameters are used, if a and x(1:100) are initialized along with the real scalar a then

call SSCAL(100,a,x(1),1) $\Rightarrow$	$x(i) \leftarrow ax(i)$	$i = 1{:}100$
call SSCAL(50,a,x(2),2) $\Rightarrow$	$x(2i) \leftarrow ax(2i)$	$i = 1{:}50$
call SSCAL(99,x(1),x(2),1) $\Rightarrow$	$x(i+1) \leftarrow x(1)\,x(i+1)$	$i = 1{:}99$

Adding a Multiple of One Vector to Another

The subroutine SAXPY is used to perform the elementary vector operation $y \leftarrow ax + y$, where a is a scalar and x and y are vectors:

```
call SAXPY( n, a, x, incx, y, incy )
```

The tops of vectors x and y are specified by x and y, and the corresponding increments by incx and incy. The effective dimension is given by n . To illustrate, if a, x(1:100), and y(1:50) are initialized then

```
call SAXPY(50,a,x,1,y,1)          ⇒ y(i) ← ax(i) + y(i)                 i = 1:50
call SAXPY(5,a,x(3),2,y(10),4)    ⇒ y(6+4i) ← ax(1+2i) + y(6+4i)        i = 1:5
call SAXPY(49,x(1),x(2),1,y,1)    ⇒ y(i) ← x(1)x(i+1) + y(i)            i = 1:49
```

Inner Products

Inner products are computed using the function SDOT. For example, the statement

$$w = \text{SDOT}(n, x, incx, y, incy)$$

computes the inner product between the two vectors x and y. The arguments x and y specify the tops of these two vectors and incx and incy specify the corresponding increments. The effective length of the inner product is passed through n . To illustrate these details, if x(1:100) and y(1:50) are initialized then

```
w = SDOT(50,x,1,y,1)      ⇒ w = x(1)y(1) + ... + x(50)y(50)
w = SDOT(50,x,2,y,1)      ⇒ w = x(1)y(1) + x(3)y(2) + ... + x(99)y(50)
w = SDOT(50,x(51),1,y,1)  ⇒ w = x(51)y(1) + ... + x(100)y(50)
w = SDOT(3,x,1,y(3),-1)   ⇒ w = x(1)y(3) + x(2)y(2) + x(3)y(1)
```

Examples follow that utilize SSCAL, SAXPY, and SDOT.

Example 2.2-1 (Univariate Least Squares)

```
          subroutine LS( m, a, b, x )
          integer m
          real a(*), b(*), x
c
c   Given m-vectors a and b, this subroutine determines a
c   scalar x such that the 2-norm of b-ax is minimized.
c
c   On entry:
c         m         integer    The dimension of vectors a and b.
c         a         real(m)    The m-vector a.
c         b         real(m)    The m-vector b.
c
c   On exit:
c
c         x         real       Returns minimizer of ||b-ax||₂
c         b         real(m)    Returns the minimum residual  b-ax.
c         a         real(m)    Returns the optimum predictor ax.
c
c   Subprograms called: SDOT, SAXPY, SSCAL
c
c   Local variables:
          real SDOT
c
c   Compute x that minimizes the 2-norm of b-ax.
c
          x   =  SDOT( m, a, 1, a, 1 )
          if ( x .NE. 0.0 )    then
             x = SDOT( m, a, 1, b, 1 )/x
c
c   Compute the minimum residual b-ax.
c
          call SAXPY( m, -x, a, 1, b, 1 )
c
c   Compute the optimum predictor ax.
c
          call SSCAL( m, x, a, 1 )
c
          endif
c
          return
          end
```

Example 2.2-2 *(Skew and Symmetric Parts)*

```
      subroutine SYMSKU( n, A, adim )
      integer n,adim
      real A(adim,*)
c
c This subroutine overwrites the upper triangular portion of
c an n-by-n matrix A with its symmetric part T = (A + A^T)/2,
c and the lower triangular portion with its skew-symmetric
c part S = (A - A^T)/2.
c
c On entry:
c       A       real(adim,n)  The matrix A.
c       adim    integer       Row dimension of the array A.
c       n       integer       Order of the matrix A.
c
c On exit:
c       A       Upper triangular part returns upper triangular
c               part of (A + A^T)/2. Strictly lower triangular
c               part returns strictly lower triangular part of
c               (A - A^T)/2.
c
c Subprograms called: SAXPY, SSCAL
c
c Local variables
c
      real half
      integer k,kp1,nmk
      parameter( half = .50e0)
c
      do 10 k = 1, n-1
         kp1 = k + 1
         nmk = n - k
         call SAXPY( nmk, 1., A(kp1,k), 1, A(k,kp1), adim)
         call SSCAL( nmk, half, A(k,kp1), adim )
         call SAXPY( nmk, -1., A(k,kp1), adim, A(kp1,k), 1 )
   10 continue
      return
      end
```

Example 2.2-3 (Matrix-Vector Multiplication)

```
          subroutine MMULT( m, n, A, adim, x, y, job )
          integer m, n, adim, job
          real A(adim,*), x(*), y(*)
c
c  This subroutine performs matrix-vector multiplication.
c  If job = 0 then y = Ax, otherwise y = A^T x.
c
c  On entry:
c       A         real(adim,n)   The matrix A.
c       adim      integer        Row dimension of array A.
c       x         real(*)        The vector x.
c       m         integer        Row dimension of matrix A,
c                                 must have m <=  adim
c       n         integer        Column dimension of matrix A.
c       job       integer        job = 0 compute y = Ax
c                                 job = 1 compute y = A^T x.
c
c  On exit:
c       y         real(*)        If job = 0, returns Ax;
c                                 If job = 1, returns A^T x
c
c  Subprograms called : SAXPY, SDOT
c
c  Local variables
          real SDOT
          integer i,j
c
          if (job .EQ. 0) then
          do 10 i = 1,m
             y(i) = 0.0
   10     continue
          do 20 j = 1,n
          call SAXPY( m, x(j), A(1,j), 1, y(1), 1)
   20     continue
          else
```

```
          do 30 j = 1,n
          y(j) = SDOT ( m, A(1,j), 1, x(1), 1 )
   30     continue
       endif
c

       return
       end
```

Example 2.2-4 (Krylov Matrix)

```
       subroutine KRYLOV( n, A, adim, K, kdim, v, j )
       integer n, adim, kdim, j
       real A(adim,*), K(kdim,*), v(*)
c
c  Given an n-by-n matrix A and an n-vector v, this
c  subroutine sets up the Krylov matrix K where
c
c                K = [ v,Av,...,A^(j-1)v ]
c
c  On entry:
c       A       real(adim,n)   The matrix A.
c       adim    integer        Row dimension of the array A.
c       v       real(n)        Contains the vector v.
c       n       integer        Dimension of matrix A and
c                              vector v. Must have n <= adim.
c       j       integer        Column dimension of matrix K.
c       kdim    integer        Row dimension of array K
c                              Must have n <= kdim.
c
c  On exit:
c       K          real(kdim,n)   Returns the Krylov matrix K.
c
c  Subprograms called:  SCOPY, MMULT
c
c  Local variables:
c          integer p
c
c  Set up the n-by-j Krylov matrix K, column by column.
```

```
c
      call SCOPY( n, v(1), 1, K(1,1), 1 )
      do 10 p = 2,j
          call MMULT( n, n, A, adim, K(1,p-1), K(1,p), 0 )
 10   continue
c

      return
      end
```

Problems for Section 2.2

1. Write a Fortran subroutine

```
      CROSS( m, n, A, adim, work )
```

that overwrites the upper triangular portion of an m-by-n matrix A with the upper triangular portion of A^TA. You will need a workspace.

2. Write a Fortran subroutine

```
      TRISQR( n, A, adim, work )
```

that overwrites an n-by-n upper triangular matrix A with A^2. Here, work is a vector workspace.

3. Write a Fortran subroutine

```
      SLVEST( n, A , adim, m, B, bdim, C, cdim )
```

that accepts as input an n-by-n matrix A and an m-by-m matrix B and returns the nm-by-nm matrix $C = I \otimes A + B \otimes I$, where "$\otimes$" denotes the Kronecker product. In particular, $(W \otimes V)$ is the block matrix $(w_{ij}V)$.

2.3 NORM COMPUTATIONS

The computation of the *p*-norm of a vector for $p = 1, 2$, or ∞ is a very frequent calculation in matrix computations. The BLAS SNRM2, SASUM, and ISAMAX are useful for this purpose.

1-Norms and 2-Norms

The 2-norm of a real *n*-vector x can be computed as follows:

$$sw = SNRM2(\; n, \; x, \; incx \;)$$

The function SASUM is used for 1-norms:

$$sw = SASUM(\; n, \; x, \; incx \;)$$

As usual, x specifies the top of the vector x, n passes the dimension, and incx the increment. To illustrate the use of these two BLAS, if x(1:20) is initialized then

```
w = SNRM2(20,x,1)      ⇒  w ← sqrt( |x(1)|² + ... + |x(20)|² )
w = SASUM(20,x,1)      ⇒  w ← |x(1)| + ... + |x(20)|
w = SNRM2(10,x(2),2)   ⇒  w ← sqrt( |x(2)|² + ... + |x(20)|² )
w = SASUM(10,x(2),2)   ⇒  w ← |x(2)| + ... + |x(20)|
```

Maximal Entry

To compute the infinity norm of a vector, one must scan for the largest component in absolute value. The BLA subprogram ISAMAX can be used for this computation. If

$$imax = ISAMAX(\; n, \; x, \; incx \;)$$

then imax contains the *index* of the component of maximal absolute value. The effective dimension of the vector scanned is passed through n. The top of the

119

vector is given by x and the increment by incx. If $x = (2, -1, 0, 6, -7)$ then

```
imax  = isamax( 5, x, 1 )        ⇒  imax ← 5
imax  = isamax( 2, x(2), 2 )     ⇒  imax ← 2
```

Example 2.3-1 (Vector Infinity-Norm)

```
      real function NORMI(x,n)
      integer n
      real x(*)
c
c  This function computes the infinity-norm of the vector x.
c  On entry:
c      x          real(n)    The vector x.
c      n          integer    Dimension of the vector x.
c
c  On exit:
c      NORMI      real    The infinity-norm of vector x.
c
c  Subprograms called: ISAMAX, abs
c
c  Local Variables
      integer ISAMAX
      real abs
c
      NORMI = abs( x( ISAMAX( n, x, 1 ) ) )
c
      return
      end
```

Example 2.3-2 (Frobenius Matrix Norm)

```
      real function SNORMF ( m, n, A, adim )
      integer m, n, adim
      real A(adim,*)
c
c  This function computes the Frobenius norm of matrix A.
c
c  On entry:
c      A       real(adim,n)   The matrix A.
c      adim    integer        Row dimension of array A.
c      m       integer        Row dimension of matrix A.
c      n       integer        Column dimension of matrix A.
c
c  On exit:
c      SNORMF  real           The Frobenius norm of matrix A.
c
c  Subprograms called: SNRM2
c
c  Local variables
      real v(2), SNRM2
      integer k
c
c  Compute the Frobenius norm of matrix A column by column,
c  updating partial result.
c
      SNORMF = 0.0
      do 10 k = 1,n
         v(1)  =  SNORMF
         v(2)  =  SNRM2( m, A(1,k), 1 )
         SNORMF  = SNRM2( 2, v(1), 1 )
 10   continue
c
      return
      end
```

Example 2.3-3 (Matrix 1-Norm)

```
        real function SNORM1( m, n, A, adim )
        integer m, n, adim
        real A(adim,*)
c
c  This function computes the 1-norm of the matrix A.
c
c  On entry:
c      A        real(adim,n)   The matrix A.
c      adim     integer        Row dimension of the array A.
c      m        integer        Row dimension of the matrix A.
c      n        integer        Column dimension of the matrix A.
c
c  On exit:
c      SNORM1 real             The 1-norm of matrix A.
c
c  Subprograms called, SASUM
c
c  Local variables
        real s, SASUM
        integer k
c
c  Compute the 1-norm of matrix A column by column, updating
c  partial result.
c
        SNORM1 = 0.0
        do 10 k = 1,n
            s = SASUM( m, A(1,k), 1 )
            SNORM1 = max( s, SNORM1 )
   10   continue
c
        return
        end
```

Example 2.3-4 (Householder Vector)

```
      subroutine HOUSE( n, x )
      integer n
      real x(*)
c
c  This subroutine overwrites an n-vector x with a
c  unit n-vector v such that if P = I-2vvᵀ, then
c  Px is  zero below the first component.
c
c  On entry:
c      x       real(n)    The vector x.
c      n       integer    Dimension of the vector x.
c
c  On exit:
c      x                  Returns the Householder vector v.
c                         Set to e₁ if x = 0.
c
c  Subprograms called.  SSCAL, SNRM2, abs
c
c  Local variables
      real alfa,beta,x1,SNRM2, abs
c
      alfa = SNRM2( n, x(1), 1 )
c
      if (alfa .NE. 0.0) then
          x1 = x(1)
          x(1) = x1 + sign(alfa,x1)
          beta = sqrt( 2.0 * alfa * (alfa + abs(x1)) )
          call SSCAL( n, 1.0/beta, x(1), 1 )
      else
          x(1) = 1.
      endif
      return
      end
```

Problems for Section 2.3

1. Write a real Fortran function

```
MAXELM( A, adim, m, n )
```

that accepts as input an m-by-n real matrix A and returns the magnitude of its largest entry.

2. Write a Fortran function

```
SORT( x, n, p )
```

that overwrites the real n-vector x with Px, where P is a permutation and the components of Px are ordered from smallest to largest in absolute value. The integer vector p should have the property that $P = [\, e_{p(1)}, ..., e_{p(n)} \,]$, where $I = [\, e_1, ..., e_n \,]$.

3. Write a Fortran subroutine

```
SEND( x, y, n )
```

that overwrites the n-vector x with a unit 2-norm n-vector v such that if $P = I - 2vv^T$ then Px is a multiple of y. Assume that x and y are nonzero.

2.4 GIVENS ROTATIONS

Two of the BLAS are concerned with Givens rotations. The subroutine SROTG constructs rotations and SROT applies them.

Constructing a Givens Rotation

The subroutine SROTG can be used to construct a Givens rotation for the purpose of zeroing the second component of a prescribed 2-vector. In particular, the call

$$\text{call SROTG(a, b, c, s)}$$

returns c and s such that

$$\begin{bmatrix} c & s \\ -s & c \end{bmatrix} \begin{bmatrix} a \\ b \end{bmatrix} = \begin{bmatrix} r \\ 0 \end{bmatrix} \qquad c^2 + s^2 = 1$$

The scalar a returns r while b returns a quantity z computed as follows:

```
if ( abs(a) .GT. abs(b) )  then
    z = s
elseif ( (abs(b) .GE. abs(a)).AND.(c .NE. 0) )  then
    z = 1/c
else
    z = 1
endif
```

It is possible to reconstruct c and s from z. This makes it possible to compactly represent the product of Givens rotations as we illustrate in Examples 2.4-1 and 2.4-2.

To see what SROTG does, consider the example $x = (1, 4, 7, 3)$. We have

```
call SROTG(x(2),x(4),c,s)   ⟹ x(2) ← 5, x(4) ← .6, c ← .8, s ← .6
call SROTG(x(4),x(2),c,s)   ⟹ x(2) ← 5/3, x(4)← 5, c ← .6, s ← .8
```

Reconstructing a Givens Rotation

The code required for the reconstruction of a Givens rotation from the single scalar encoding z is encapsulated in the following subroutine.

```
      subroutine ZTOCS( z, c, s )
      real z,c,s
c
c  Computes a sine-cosine pair from scalar  z  produced by
c  SROTG.
c
c  Subprograms called: abs, sqrt
c
      if ( abs(z) .LT. 1 ) then
          s = z
          c = sqrt(1. - s*s)
      elseif ( abs(z) .GT. 1 ) then
          c = 1./z
          s = sqrt(1. - c*c)
      else
          c = 0.
          s = 1.
      endif
c
      return
      end
```

Applying a Givens Rotation

The subroutine SROT is used when a Givens rotation must be applied to a pair of vectors. In particular,

```
      call SROT( n, x, incx, y, incy, c, s )
```

prompts the computation

$$[x \quad y] \leftarrow [x \quad y] \begin{bmatrix} c & -s \\ s & c \end{bmatrix}$$

As usual, in the calling sequence x and y are one-dimensional arrays, n is the dimension of the vectors x and y that they house, and incx and incy the respective increments. In a typical application of SROT, x and y refer to a pair of matrix rows or columns. To illustrate, let the notation $G = J(p,q,c,s)$ imply that G is the identity matrix except that $g_{pp} = g_{qq} = c$ and $g_{pq} = -g_{qp} = s$, i.e., G is a Givens rotation. Suppose A is an m-by-n matrix sitting in an array A that has row dimension adim. Then

call SROT(n,A(p,1),adim,A(q,1),adim,c,s) $\Rightarrow$ $A = J(p,q,c,s)A$

call SROT(n,A(p,1),adim,A(q,1),adim,c,-s) $\Rightarrow$ $A = J(p,q,c,s)^T A$

call SROT(m,A(1,p),1,A(1,q),1,c,s) $\Rightarrow$ $A = A\, J(p,q,c,s)$

call SROT(m,A(1,p),1,A(1,q),1,c,-s) $\Rightarrow$ $A = A\, J(p,q,c,s)^T$

Example 2.4-1 (Zeroing an n-Vector)

```
      subroutine ZERO(n,x)
      integer n
      real x(*)
c
c  This subroutine computes an orthogonal matrix Q such that
c  Q*x is a multiple of the first column of the identity I.
c  Q = G(1)...G(n-1), where G(k) is a Givens rotation in
c  plane (k,k+1). Information necessary to reconstruct G(k)
c  is returned in x(k+1). See APPLY in next example.
c
c  On entry :
c     x        real(n)      The vector x.
c     n        integer      Dimension of vector x.
c
c  On exit:
c     x        Returns the information necessary to
c              reconstruct  G(k) for k = 1,...,n-1. Used by
c              APPLY.
c
c  Subprograms called:  SROTG
c
c     Local variables
      real c,s
      integer k
c
      do 10 k = n-1,1,-1
c
c  Zero x(k+1) and record the Givens rotation.
c
         call SROTG(x(k),x(k+1),c,s)
   10 continue
c
      return
      end
```

Example 2.4-2 (Factored Q Times Vector)

```
      subroutine APPLY( n, x, y )
      integer n
      real x(*), y(*)
c
c   This subroutine overwrites the vector y with the vector
c   Qy, where Q = G(1) ... G(n-1) is an orthogonal matrix
c   produced by the subroutine ZERO.
c
c   On entry:
c       x   real(n)    The vector x from subroutine ZERO.
c       y   real(n)    The vector y.
c       n   integer    Dimension of the vectors x and y.
c
c   On exit:
c       y               Returns the vector Qy.
c
c   Subprograms called: SROT, ZTOCS
c
c   Local variables
      real c, s
      integer k
c
c   Reconstruct both c and s from z, and apply Givens
c   rotation to the vector y.
c
      do 10 k = n-1, 1, -1
            call ZTOCS( x(k+1), c, s )
            call SROT( 1, y(k), 1, y(k+1), 1, c, s )
   10 continue
c
      return
      end
```

Example 2.4-3 (*Hessenberg QR Factorization*)

```
      subroutine HESS( n, A, adim )
      integer n, adim
      real A(adim,*)
c
c  This subroutine overwrites the upper triangular portion of
c  the n-by-n upper Hessenberg matrix A with the
c  upper triangular matrix Q^T A = R, where Q is orthogonal.
c
c  On entry:
c      A       real(adim,n)      The  Hessenberg matrix A.
c      adim    integer           Row dimension of the array A.
c      n       integer           Order of Hessenberg matrix A.
c
c  On exit:
c      A          Returns the upper triangular matrix Q^T A.
c
c  Subprograms called: SROT, SROTG
c
c  Local variables:
      real c, s
      integer j, jp1
c
c  Compute Q = G(1)...G(n-1) where Q is orthogonal, and
c  apply rotations to A successively to compute  R = Q^T A.
c
      do 20 j = 1, n-1
         jp1 = j + 1
         call SROTG( A(j,j), A(jp1,j), c, s )
         call SROT( n-j, A(j,jp1), adim, A(jp1,jp1),
     &                             adim, c, s)
   20 continue
c
      return
      end
```

Problems for Section 2.4

1. Write a Fortran subroutine

```
MAP ( x, y, c, s )
```

that accepts real 2-vectors x and y (both nonzero) and computes a Givens rotation G such that Gx is a multiple of y.

2. Write a Fortran subroutine

```
SIMIL( A, adim, n, i, j, c, s )
```

that overwrites the n-by-n real matrix A with $Q^T A Q$, where Q is the identity everywhere except $q_{ii} = q_{jj} = c$ and $q_{ij} = -q_{ji} = s$.

3. Write a Fortran subroutine

```
SYM2 ( B, bdim, c, s )
```

that accepts as input a real 2-by-2 matrix B and computes a Givens rotation G such that the product GB is symmetric.

2.5 DOUBLE PRECISION AND COMPLEX VERSIONS

With LINPACK it is possible to use real or complex arithmetic with single or double precision. A system of prefixes is used to designate which of the four types prevails in a given subprogram:

      s   real single precision
      d   real double precision
      c   complex single precision
      z   complex double precision

In this section we discuss these options as they pertain to the BLAS. To facilitate the discussion, the above prefixes are also used to designate the type of an argument. Thus,

```
dw  =  DZNRM2( n, zx, incx )
```

computes the 2-norm of a double precision complex vector. The subprogram is called DZNRM2 rather than ZNRM2 because it returns a real double precision result.

Because it is not strictly a part of Fortran 77, we mention that complex double precision variables are declared as follows:

complex*16 { *list-of-variables* }

Copying a Vector

```
call  SCOPY( n, sx, incx, sy, incy )
call  DCOPY( n, dx, incx, dy, incy )
call  CCOPY( n, cx, incx, cy, incy )
call  ZCOPY( n, zx, incy, zy, incy )
```

Interchanging Vectors

```
call   SSWAP( n, sx, incx, sy, incy )
call   DSWAP( n, dx, incx, dy, incy )
call   CSWAP( n, cx, incx, cy, incy )
call   ZSWAP( n, zx, incx, zy, incy )
```

Multiplying a Vector by a Scalar

```
call SSCAL( n, sa, sx, incx )
call DSCAL( n, da, dx, incx )
call CSCAL( n, ca, cx, incx )
call ZSCAL( n, za, zx, incx )
```

In addition, there are BLAS for multiplying complex vectors by real scalars:

```
call   CSSCAL( n, sa, cx, incx )
call   ZDSCAL( n, da, zx, incx )
```

Inner Product

```
sw  =  SDOT( n, sx, incx, sy, incy )
dw  =  DDOT( n, dx, incx, dy, incy )
```

For the complex case, there are BLAS for $x^H y$,

```
cw =  CDOTC( n, cx, incx, cy, incy )
zw =  ZDOTC( n, zx, incx, zy, incy )
```

and for $x^T y$,

```
cw =  CDOTU( n, cx, incx, cy, incy )
zw =  ZDOTU( n, cx, incx, cy, incy )
```

The "c" and "u" suffixes stand for *conjugate* and *unconjugate*.

Adding a Multiple of One Vector to Another

```
call  SAXPY( n, sa, sx, incx, sy, incy )
call  DAXPY( n, da, dx, incx, dy, incy )
call  CAXPY( n, ca, cx, incx, cy, incy )
call  ZAXPY( n, za, zx, incx, zy, incy )
```

2-Norm

```
sw  =  SNRM2 ( n, sx, incx )
dw  =  DNRM2 ( n, dx, incx )
sw  =  SCNRM2( n, cx, incx )
dw  =  DZNRM2( n, zx, incx )
```

1-Norm

```
sw  =  SASUM ( n, sx, incx )
dw  =  DASUM ( n, dx, incx )
sw  =  SCASUM( n, dx, incx )
dw  =  DZASUM( n, zx, incx )
```

In the complex case, the quantity $||Re(x)||_1 + ||Im(x)||_1$ is returned rather than the actual 1-norm.

Maximal Entry

```
imax  =  ISAMAX( n, sx, incx )
imax  =  IDAMAX( n, dx, incx )
imax  =  ICAMAX( n, cx, incx )
imax  =  IZAMAX( n, zx, incx )
```

In the complex cases, the smallest k that maximizes the quantity $|Re(x(k))| + |Im(x(k))|$ is returned.

Constructing a Givens Rotation

```
call SROTG( sa, sb, sc, ss )
call DROTG( da, db, dc, ds )
```

There is no BLA routine for constructing complex Givens rotations. (However, see the example below.)

Applying a Givens Rotation

```
call  SROT ( n, sx, incx, sy, incy, sc, ss )
call  DROT ( n, dx, incx, dy, incy, dc, ds )
call  CSROT( n, cx, incx, cy, incy, sc, ss )
call  ZDROT( n, zx, incx, zy, incy, dc, ds )
```

Note that `CSROT` and `ZDROT` apply real Givens rotations to complex vectors.

Example 2.5-1 (Complex Rotations)

```
      subroutine SCROTG( ca, cb, sc1, ss1, sc2, ss2 )
      complex ca,cb
      real sc1,ss1,sc2,ss2
c
c  Given complex numbers    ca = a1 + i*a2
c                           cb = b1 + i*b2
c
c  This subroutine computes real cosine-sine pairs
c  (c1,s1) and (c2,s2) so
c
c  [    c1        s1(c2+i*s2) ] [ a1 + i*a2 ] = [ r ]
c  [-s1(c2-i*s2)      c1      ] [ b1+  i*b2 ]   [ 0 ]
c
c  On entry :
c      ca      complex    1st component of vector to be zeroed.
```

```
c       cb       complex    2nd component of vector to be zeroed.
c
c  On exit:
c       sc1      real       c1
c       ss1      real       s1
c       sc2      real       c2
c       ss2      real       s2
c
c  Subprograms called: SROTG, real, aimag
c
c  Local variables
        real a1, a2, b1, b2
        real real, aimag
c
c  Compute real and imaginary parts of ca and cb,
c  the 1st and 2nd component of vector to be zeroed.
c
        a1  =  real(ca)
        a2  =  aimag(ca)
        b1  =  real(cb)
        b2  =  aimag(cb)
c


c     Compute polar representations of ca and cb.
c
        call SROTG( a1, a2, sca, ssa )
        call SROTG( b1, b2, scb, ssb )
c
c     Compute the cosine-sine pairs (sc1,ss1) and (sc2,ss2).
c
        call SROTG( a1, b1, sc1, ss1 )
        sc2  =  sca*scb + ssa*ssb
        ss2  =  ssa*scb - sca*ssb
c
        return
        end
```

Problems for Section 2.5

1. Devise a scheme whereby the subroutine SCROTG in Example 2.5-1 encodes the information necessary to reconstruct the complex Givens rotation in a single complex variable.

2. Write a subroutine HOUSE(x,n) that overwrites the complex n-vector x with a complex unit n-vector v such that if $P = I - 2vv^H$ then Px is a multiple of the first column of the identity.

3. Write a complex single precision version of the subroutine LS in Example 2.2-1.

4. Write a complex double precision version of the subroutine MMULT in Example 2.2-2.

5. Write a complex double precision version of the subroutine SNORMF in Example 2.3-2.

Chapter 3

LINPACK

LINPACK is a collection of Fortran subroutines that can be used to solve numerous linear system and least squares problems. For square systems it has routines that perform the following factorizations:

$$PA = LU \quad (A \text{ general, full, or banded})$$
$$A = GG^T \quad (A \text{ symmetric positive definite, full, or banded})$$
$$A = LDL^T \quad (A \text{ symmetric})$$

Depending upon the choice of subroutine, an estimate of A's condition number may attend this phase of the calculation. Following factorization, a second subroutine must typically be called to produce a solution to $Ax = b$.

With LINPACK you can also compute the following orthogonal matrix factorizations:

$$AP = QR \quad (\text{Orthonormalization})$$
$$A = USV^T \quad (\text{Singular Value Decomposition})$$

These factorizations can be used to solve overdetermined least squares problems.

This chapter is organized according to the type of problem solved: triangular systems in §3.1, general linear systems in §3.2, definite and indefinite symmetric systems in §3.3, and banded systems in §3.4. The QR and singular value decompositions are then discussed in §§3.5 and 3.6, with emphasis on least squares problems. LINPACK codes are implemented in four different precisions. We detail the real single precision versions in §§3.1-3.6. The complex and double precision variants are discussed in §3.7.

Further information on LINPACK is available in the manual [1].

3.1 TRIANGULAR SYSTEMS

The standard approach to solving dense linear systems reduces the given problem to a collection of triangular problems. Thus, we begin with the LINPACK codes associated with the solution of upper or lower triangular systems.

Solving Triangular Systems

The subroutine STRSL can be used to solve systems of the form

$$Tx = b$$

and

$$T^T x = b$$

where T is either upper triangular or lower triangular. A call to STRSL has the form

```
call STRSL( T, tdim, n, b, job, info )
```

where

T	real (tdim,n)	The triangular matrix T.
tdim	integer	The row dimension of the array T.
n	integer	The dimension of the matrix T.
b	real (n)	The right-hand side. Returns $T^{-1}b$ if T is nonsingular. Otherwise unchanged.
job	integer	Indicates the type of system to be solved.
info	integer	Returns a singularity flag.

The argument job is encoded as follows:

$$
\begin{array}{llll}
\texttt{job} = 00 & \Rightarrow & \text{Solve } Tx = b, & \text{where } T \text{ is lower triangular} \\
\texttt{job} = 01 & \Rightarrow & \text{Solve } Tx = b, & \text{where } T \text{ is upper triangular} \\
\texttt{job} = 10 & \Rightarrow & \text{Solve } T^T x = b, & \text{where } T \text{ is lower triangular} \\
\texttt{job} = 11 & \Rightarrow & \text{Solve } T^T x = b, & \text{where } T \text{ is upper triangular}
\end{array}
$$

The variable info is used to communicate singularity. If T is nonsingular, then upon return info is zero. If T is singular then info returns the smallest k such that t_{kk} is zero. The array T itself is unaltered by STRSL.

Condition Estimation

The 1-norm condition $\kappa_1(T)$ of a triangular matrix can be estimated using the subroutine STRCO :

```
call STRCO( T, tdim, n, rcond, z, job )
```

Here,

T	real (tdim,n)	The triangular matrix T.
tdim	integer	The row dimension of the array T.
n	integer	The dimension of the matrix T.
rcond	real	Returns estimate of $1/\kappa_1(T)$.
z	real (n)	Returns unit 1-norm z so $\lVert Tz \rVert_1 = rcond \lVert z \rVert_1$.
job	integer	Indicates whether T is upper or lower triangular ($0 = $ lower, $1 = $ upper) .

If the value returned by rcond is sufficiently small, then one may conclude that T is numerically singular . This is because

$$\| Tz \|_1 = rcond \| z \|_1 \approx \min_{\| x \|_1 = 1} \| Tx \|_1$$

The definition of "sufficiently small" usually means that the value is so small that we can make the stored T exactly singular by altering least significant mantissa bits. Such a decision usually affects subsequent computation in a serious way. To check for this, code of the following form should be executed:

```
rcond1 = rcond + 1.0e0
if (rcond1 .GT. 1.0e0 )    then
     { Process nonsingular T case }
else
     { Process the singular T case }
endif
```

The idea behind this kind of testing is detailed in §1.9. Essentially, if $rcond1$ is 1, then $rcond$ is less than the machine precision.

Condition estimation has a prominent role to play in the subroutines in this chapter that are concerned with linear system solving. Indeed, we stress the use of LINPACK routines that return condition estimates. That way, when a computed solution x_c to $Ax = b$ is obtained then the heuristic

$$rcond \| x - x_c \|_1 \approx \| x \|_1$$

can be invoked.

Determinant and Explicit Inverse

The explicit inverse and determinant of a triangular matrix can be computed with the subroutine STRDI:

```
call STRDI( T, tdim, n, d, job, info )
```

On input the variable job contains a three-digit integer ABC that is used to indicate the required computations. The argument d is a two-vector.

$$A = 1 \quad \Rightarrow \quad det(T) = d(1) \times 10^{d(2)}$$
$$A = 0 \quad \Rightarrow \quad \text{No determinant computations}$$
$$B = 1 \quad \Rightarrow \quad \text{Overwrite } T \text{ with its inverse}$$
$$B = 0 \quad \Rightarrow \quad \text{No inverse computations}$$
$$C = 1 \quad \Rightarrow \quad T \text{ is upper triangular}$$
$$C = 0 \quad \Rightarrow \quad T \text{ is lower triangular}$$

Example 3.1-1 (Multiple Right-hand Sides)

```
      subroutine TINVB( T,tdim,n,B,bdim,p,info,rcond,z )
c
      integer tdim, n, bdim, p, info
      real T(tdim,*), B(bdim,*), rcond, z(*)


c
c Overwrites B (nxp) with solution to TX = B where T (nxn)
c is upper triangular. If T is numerically singular,
c however, then no attempt is made to compute  X.
c
c On entry:
c     T        real(tdim,n)  The matrix   T.
c     tdim     integer       Row dimension of the array T.
c     n        integer       Dimension of the matrix   T.
c     B        real(bdim,p)  The matrix   B.
c     bdim     integer       The row dimension of array B.
c     p        integer       The column dimension of matrix B.
c
c On exit:
c     B        real(bdim,p)  Returns solution to TX= B
c                            if T nonsingular.
c     info     integer       If info = 0, T nonsingular.
c                            If info = 1, T numerically
c                            singular.
c     rcond    real          Returns 1/cond_1( T ) estimate.
c     z        real (n)      Returns  unit 1-norm z  so
c                            || Tz || = rcond|| z || in 1-norm
```

```
c
c   Local Variables
c
      integer j
      real rcond1
c
c   Uses LINPACK routines STRCO and STRSL.
c
      call STRCO( T, tdim, n, rcond, z, 1 )
      rcond1 = 1.0e0 + rcond
      if ( rcond1 .EQ. 1.0e0 ) then
         info = 1
      else
         info = 0
         do 10 j = 1, p
            call STRSL( T, tdim, n, B(1,j), 01, info )
 10      continue
      endif
      return
      end
```

Problems for Section 3.1

1. Write a subroutine

```
      TRANK1( T, tdim, n, u, v, b, info )
```

that overwrites b with the solution to $(T + uv^T)x = b$. Here, T is an n-by-n upper triangular matrix and u, v, and b are n-vectors. The integer variable info should communicate trouble encountered during the solution process.

2. Change the subroutine in Example 3.1-1 so that if B is a p-by-n matrix it overwrites B with the solution to $XA = B$.

3. Write a subroutine

```
      STRINV( T, tdim, n, R, rdim, job, info )
```

that accepts as input two n-by-n upper triangular matrices T and R, and overwrites T with TR^{-1} if $job = 0$, and with $R^{-1}T$ if $job = 1$. The integer variable info should be used to communicate trouble encountered in the solution process.

4. Write a subroutine

```
TRILYP( U, udim, n, C, cdim, work, info )
```

that solves the Lyapunov system $UX + XU^T = C$, where U is an n-by-n lower triangular matrix and C is an n-by-n symmetric matrix. Overwrite C with the solution and use info as a trouble indicator.

3.2 GENERAL SYSTEMS

There are two phases in dense linear system solving. First, the matrix is factored and then the resulting triangular systems must be solved. LINPACK provides codes for each of these phases.

LU Factorization

The subroutine SGECO computes the factorization $PA = LU$ via Gaussian elimination with partial pivoting. It should be used for general square matrices. It also provides an estimate of the condition. A call to SGECO has the form

```
CALL SGECO( A, adim, n, ipvt, rcond, z )
```

where

A	real (adim, n)	Contains the n-by-n matrix A. Returns L and U.
adim	integer	The row dimension of the array A.
n	integer	The dimension of the matrix A.
rcond	real	Returns estimate of $1/\kappa_1(A)$.
ipvt	integer (n)	Returns encoding of permutation P.
z	real (n)	Returns unit 1-norm z so $\|Az\|_1 = rcond\|z\|_1$.

Gauss transforms $M_1, \ldots, M_{n-1}$ and permutations $P_1, \ldots, P_{n-1}$ are generated so that $M_{n-1}P_{n-1}\cdots M_1P_1A = U$ is upper triangular. The upper triangular portion of A is overwritten by U, whereas for all $i > k$, $A(i,k)$ is overwritten by the (i,k) entry of M_k. The permutation P_k can be obtained by interchanging rows k and $ipvt(k)$ of I_n .

If the condition number is not required, then use SGEFA instead of SGECO:

$$\text{CALL SGEFA(A, adim, n, ipvt, info)}$$

All the arguments have the same role as in SGECO except the integer variable info which is used to communicate trouble in the event of ill-conditioning:

$$\text{info} = 0 \quad \Rightarrow \quad \text{No zero pivots encountered}$$

$$\text{info} = k \quad \Rightarrow \quad u_{kk} = 0$$

Because it does a better job of spotting numerical singularity, SGECO should be used instead of SGEFA whenever ill-conditioning is suspected.

Back-substitution/Forward-elimination

The output of SGECO or SGEFA can be used by the subroutine SGESL to solve either the system $Ax = b$ or $A^T x = b$. A call to SGESL has the form

$$\text{CALL SGESL(A, adim, n, ipvt, b, job)}$$

where

A	real (adim,n)	Contains A's LU decomposition from either SGECO or SGEFA .
adim	integer	The row dimension of the array A.
n	integer	The dimension of the matrix A .
ipvt	integer (n)	The pivot vector from SGECO.
b	real (n)	The right-hand side. Returns solution to requested system.

| `job` | integer | $job = 0 \implies$ solve $Ax = b$ |
| | | $job = 1 \implies$ solve $A^T x = b$ |

Explicit Inverse and Determinant

Once either `SGECO` or `SGEFA` has been called, then the explicit inverse and determinant of A can be calculated by using the subroutine `SGEDI`:

```
call SGEDI( A, adim, n, ipvt, d, work, job )
```

The integer variable `job` indicates what is to be done:

`job = 01`	$\implies$	Only inverse
`job = 10`	$\implies$	Only determinant
`job = 11`	$\implies$	Determinant and inverse

A real workspace `work(n)` is required. When the inverse is formed, it overwrites A. The determinant is returned in a two-vector `det` that satisfies $det(A) = d(1) \, 10^{d(2)}$.

Example 3.2-1 (Ax = b Solver with Condition Estimate)

```
      subroutine SGSOL( A, adim, n, ipvt, rcond, z, b, info)
c
      integer adim , n,  ipvt(*), info
      real A(adim,*),  b(*),  rcond,  z(*)
c
c  On entry:
c      A       real(adim,n)  The matrix A.
c      adim    integer       The row dimension of the array A.
c      n       integer       The dimension of the matrix A.
c                            Must have n <= adim.
c      b       real(n)       The right-hand side vector.
c
c  On exit:
c      b       real(n)       Returns A⁻¹b if A is
c                            not numerically singular.
c      ipvt    integer(n)    Used for pivot information.
c      rcond   real          Returns 1/cond₁( A ) estimate.
c      z       real (n)      Returns  unit 1-norm z  so
c                            || Az || = rcond|| z || in 1-norm
c      info    integer       info = 0  ⟹  x found
c                            info > 0  ⟹  A singular, no x
c
c  Uses LINPACK routines SGECO and SGESL.
c  Local Variables:
c
      real rcond1
c
      call SGECO ( A, adim,  n, ipvt, rcond, b )
      rcond1 = rcond + 1.0e0
      if ( rcond1  .GT.  1.0e0 )    then
         info = 0
         call SGESL ( A, adim, n, ipvt, b, 0 )
      else
         info = 1
      endif
c
      return
      end
```

It is important to recall that near singularity in A can result in overflow in SGESL. That is why the condition of A is checked before SGESL can be invoked.

Problems for Section 3.2

1. Change the subroutine in Example 3.1-1 so that it handles the case when T is full instead of upper triangular.

2. Write a subroutine

```
BILIN( A, adim, n, u, v, f, info )
```

that computes $f = u^T A^{-1} v$, where u and v are real n-vectors and A is an n-by-n real nonsingular matrix. The integer variable `info` should flag situations where A is numerically singular.

3. Write a subroutine

```
EMAX( A, adim, n, ipvt, w )
```

that takes the output from SGECO (i.e., the factorization $PA = LU$) and computes the real number

$$w = ||\ |\ U^T|\ |L^T|\ e||\ _\infty$$

where e is the vector of all 1's. Note: the absolute value of a matrix is obtained by taking the absolute value of its entries.

3.3 SYMMETRIC SYSTEMS

In this section the subroutines described are concerned with full symmetric matrices. In the positive definite case, these routines involve the computation of the Cholesky factorization. In the indefinite case, the method of diagonal pivoting is used. In both situations there are routines for when A is stored conventionally and for when it is stored in "packed" form.

Cholesky Factorization

The subroutine SPOCO can be used to compute the Cholesky factorization $A = R^T R$ of a symmetric positive definite (s.p.d.) matrix :

```
CALL SPOCO( A, adim, n, rcond, z, info )
```

Here:

A	real (adim,n)	Contains the n-by-n s.p.d. matrix A. Returns R in the upper triangular portion of the matrix A, where $A = R^T R$.
adim	integer	The row dimension of the array A.
n	integer	The dimension of the matrix A. Must have $n \le adim$.
rcond	real	Returns $1/\kappa_1(A)$ estimate.
z	real (n)	Returns unit 1-norm z so $\|Az\|_1 = rcond\|z\|_1$.
k	integer	$k = 0 \implies R$ is found. $k > 0 \implies A(1{:}k,1{:}k)$ is not p.d.

If condition estimation is not required, then

```
SPOFA( A, adim, n, info )
```

can be used to compute the Cholesky factorization. The arguments perform the same function as in SPOCO.

The output of either SPOCO or SPOFA can be used by the subroutine SPOSL to solve a symmetric positive definite system $Ax = b$:

```
CALL SPOSL( A, adim, n, b )
```

Here,

A	real (adim,n)	A's Cholesky factor produced by either SPOCO or SPOFA.
adim	integer	Row dimension of the array A.
n	integer	Dimension of the matrix A.
b	real (n)	The right-hand side. Returns $A^{-1}b$.

Diagonal Pivoting Method

Symmetric indefinite systems can be solved using the subroutines SSICO and SSISL. The first of these computes the factorization $P^T A P = U D U^T$, where D is block diagonal with 1-by-1 and 2-by-2 blocks and where P is a permutation chosen so that the upper triangular matrix U has entries less than 1 in magnitude:

```
CALL SSICO( A, adim, n, ipvt, rcond, z )
```

Here,

A	real (adim,n)	The symmetric A. Returns U and D.
adim	integer	The row dimension of the array A.
n	integer	The dimension of the matrix A. Must have $n \le adim$.

ipvt	integer(n)	Returns pivot information.
rcond	real	Returns $1/\kappa_1(A)$ estimate.
z	real (n)	Returns unit 1-norm z so $\|Az\|_1 = rcond\|z\|_1$.

If the condition estimate is not desired then use

$$\text{CALL SSIFA(A, adim, n, ipvt, k)}$$

The integer variable k communicates singularity:

$$k = 0 \qquad \Rightarrow \quad A \text{ nonsingular}$$

$$k > 0 \qquad \Rightarrow \quad k\text{th diagonal block of } D \text{ is singular}$$

The other arguments have exactly the same function as in SSICO.

The output of either SSICO or SSIFA can be used to solve a symmetric indefinite system $Ax = b$ by invoking SSISL:

$$\text{CALL SSISL(A, adim, n, ipvt, b)}$$

Here,

A	real (adim,n)	Factored A from SSICO or SSIFA.
adim	integer	The row dimension of the array A.
n	integer	The dimension of the matrix A. Must have $n \leq adim$.
ipvt	integer (n)	Pivot information from SSICO or SSIFA.
b	real (n)	The right-hand side. Returns $A^{-1}b$.

Inverses, Determinant, Inertia

After the Cholesky factorization has been calculated via SPOCO or SPOFA, then the explicit inverse of a symmetric positive definite matrix and/or its determinant can be found using SPODI:

```
CALL SPODI( A, adim, n, det , job )
```

The integer variable job is used to prescribe whether the inverse and/or determinant is required.

In the indefinite case, the subroutine SSIDI serves a similar role:

```
CALL SSIDI( A, adim, n, kpvt, det, inert, work, job )
```

SSIDI must be called after A has been processed by SSICO or SSIFA. If the inverse is required, a workspace work(n) is needed. The inertia of A is returned in an integer three-vector:

$$
\begin{aligned}
inert(1) &= \text{the number of positive eigenvalues} \\
inert(2) &= \text{the number of zero eigenvalues} \\
inert(3) &= \text{the number of negative eigenvalues}
\end{aligned}
$$

Packed Storage

In some applications it is convenient to store symmetric matrices in "packed" form. This means that the upper triangular part of the symmetric matrix A is stored column by column in a linear array AP having dimension at least $n(n+1)/2$ (see §1.5). Thus, for $i < j$, the (i,j) entry of the matrix A is housed in AP(i+j(j-1)/2). All of the subroutines mentioned in this section have packed storage implementations in LINPACK, e.g.,

Packed Version	Conventional Version
SPPCO(AP,n,rcond,z,info)	SPOCO(A,adim,n,rcond,z,info)
SPPSL(AP,n,b)	SPOSL(A,adim,n,b)
SSPCO(AP,n,ipvt,rcond,z)	SSICO(A,adim,n,ipvt,rcond,z)
SSPSL(AP,n,ipvt,b)	SSISL(A,adim,n,ipvt,b)

Packed-storage versions of SPOFA, SPODI , SSIFA, and SSIFI also exist.

Cholesky with Pivoting

The decomposition $PAP^T = GG^T$, where A is symmetric positive semidefinite and P is a permutation matrix that can be computed using the subroutine SCHDC :

```
CALL SCHDC( A, adim, n, work, ipvt, job, info )
```

The determination of P is dictated by the integer variable ipvt(n). Roughly speaking, among the "free" variables designated by ipvt, the algorithm moves the largest "remaining" diagonal entry into the pivot position. The permutation is encoded in ipvt upon exit.

Updating and Downdating the Cholesky Factorization

Suppose that the Cholesky factorization $A = GG^T$ is known and that x is a given n-vector. The subroutine SCHUD can be used to obtain the Cholesky factorization of

$$A_+ = A + xx^T$$

while SCHDD can be used to compute the Cholesky factor (if it exists) of

$$A_- = A - xx^T$$

The subroutine SCHEX is frequently used in conjunction with SCHUD and SCHDD. It performs symmetric permutations of the form $A \leftarrow PAP^T$.

Example 3.3-1 (*Symmetric Positive Definite System Solver*)

```
      subroutine SPOSOL( A, adim, n, b, rcond, z, info )
c
      integer  n, adim, info
      real  A(adim,*), b(*), rcond, z(*)
c
c  Solves Ax = b, where A is symmetric and positive definite.
c
c  On entry:
c      A        real(adim,n)  Contains the s.p.d. matrix  A.
c      adim     integer       Row dimension of the array A.
c      n        integer       Dimension of the matrix  A.
c      b        real(n)       The right-hand side.
c
c  On exit:
c      b        real(n)       Usually returns solution to Ax = b.
c      rcond    real          Returns 1/cond_1( A ) estimate.
c      z        real (n)      Returns  unit 1-norm  z  so
c                             || Az || = rcond|| z || in 1-norm.
c      info     integer       info = 0  => b returns A^-1 b.
c                             info > 0  => A(1:k,1:k) not pos.def.
c                             info = -1 => A numerically singular.
c  Uses LINPACK routines SPOCO and SPOSL
c  Local Variables
c
      real rcond1
c
      call SPOCO( A, adim, n, rcond, z, info )
      rcond1 = rcond + 1.0e0
      if ( info .EQ. 0  .AND. rcond1 .GT. 1.0e0 ) then
         call SPOSL( A, adim, n, b )
      elseif( info .EQ. 0  .AND. rcond1 .EQ. 1.0e0 ) then
         info = -1
      endif
c
      return
      end
```

Example 3.3-2 *(Symmetric Indefinite System Solver)*

```
subroutine SSISOL( A,adim,n,ipvt,rcond,z,b,info )
integer adim, n, ipvt(*), info
real A(adim,*), b(*), rcond, z(*)
```

```
c
c  Solves  Ax = b  where  A  is symmetric.
c
c  On entry:
c      A       real(adim,n)   The symmetric matrix  A.
c      adim    integer        Row dimension of the array A.
c      n       integer        Dimension of the matrix  A.
c      ipvt    integer(n)     Pivot vector information.
c      b       real(n)        The right-hand side.
c
c  On exit:
c      b       real(n)        Usually returns A⁻¹b.
c      A       real(adim,n)   Upper triangular part returns U.
c      rcond   real           Returns 1/cond₁( A ) estimate.
c      z       real (n)       Returns  unit 1-norm z so
c                             ‖ Az ‖ = rcond‖ z ‖ in 1-norm.
c      info    integer        info = 0  ⇒ x found.
c                             info = -1 ⇒ A numerically
c                                              singular.
c
c  Uses LINPACK routines SSISL and SSICO.
c  Local Variables:
        real rcond1
c
        call SSICO( A, adim, n, ipvt, rcond, z )
        rcond1 = 1.0e0 + rcond
        if (rcond1 .GT. 1.0e0) then
            info = 0
            call SSISL( A, adim, n, ipvt, b )
        else
            info = -1
        endif
        return
        end
```

Problems for Section 3.3

1. Write a subroutine

```
FORM( A, adim, n, ipvt, b, rcond, z, w, info )
```

that computes the quantity $w = b^T A^{-1} b$, where A is symmetric and b is a given n-vector. The integer variable `info` should indicate if A is numerically singular.

2. Write a subroutine

```
LR( A, adim, n, info, work )
```

that overwrites the upper triangular part of A with the upper triangular part of RR^T, where $A = R^T R$ is A's Cholesky factorization.

3. Let A be symmetric positive definite. Write a subroutine

```
UUT( A, adim, n, info )
```

that overwrites the upper triangular part of A with an upper triangular matrix U such that $A = UU^T$. Hint: Apply `SPOFA` to the matrix EAE, where E is the exchange matrix, i.e., the identity with the column ordering reversed.

4. Let A be an n-by-n symmetric positive definite matrix and let $A_{[i,j]}$ denote the principal submatrix defined by rows and columns $i ,..., j$. Assume that $1 \le i < j \le n$. Write a subroutine

```
AIJINV( A, adim, n, i, j, info, ... )
```

that overwrites the upper triangular portion of $A_{[i,j]}$ with the upper triangular portion of its inverse.

5. Let A and B be n-by-n symmetric matrices with B positive definite. Write a subroutine

```
CONVERT( A, adim, n, B, bdim, info, work )
```

that overwrites the upper triangular part of A with the upper triangular part of $R^{-T}AR^{-1}$, where $B = R^TR$ is the Cholesky factorization of B.

3.4 BANDED SYSTEMS

Many of the subroutines in the previous sections have counterparts when the matrix A is banded. The notions of upper and lower bandwidth must be clarified before we can proceed. We say that A has *lower bandwidth p* if

$$i > j + p \quad \Rightarrow \quad a_{ij} = 0$$

and *upper bandwidth q* if

$$i < j - q \quad \Rightarrow \quad a_{ij} = 0$$

Band Gaussian Elimination

Subroutines for full general matrices were discussed in §3.2. Here are the corresponding banded versions:

```
Full:     SGECO( A, adim, n, ipvt, rcond, z )
Banded:   SGBCO( A, adim, n, p, q, ipvt, rcond, z )

Full:     SGEFA( A, adim, n, ipvt, info )
Banded:   SGBFA( A, adim, n, p, q, ipvt, info )

Full:     SGESL( A, adim, n, ipvt, b, job )
Banded:   SGBSL( A, adim, n, p, q, ipvt, b, job )

Full:     SGEDI( A, adim, n, ipvt, b, job )
Banded:   SGBDI( A, adim, n, p, q, ipvt, b, det )
```

The arguments to the band routines perform the same function as their counterparts in the full routines. The integer variables p and q pass the lower and upper bandwidths. However, the n-by-n matrix A is *not* stored conventionally. Instead, it is stored in an array of size $(2p + q + 1)$-by-n. The case $n = 7$, $p = 1$, and $q = 3$ amply illustrates how this is done:

$$
A \;=\; \begin{bmatrix}
x & x & x & a_{14} & a_{25} & a_{36} & a_{47} & a_{58} \\
x & x & a_{13} & a_{24} & a_{35} & a_{46} & a_{57} & a_{68} \\
x & a_{12} & a_{23} & a_{34} & a_{45} & a_{56} & a_{67} & a_{78} \\
a_{11} & a_{22} & a_{33} & a_{44} & a_{55} & a_{66} & a_{77} & a_{88} \\
a_{21} & a_{32} & a_{43} & a_{54} & a_{65} & a_{76} & a_{87} & x
\end{bmatrix}
$$

The general strategy is to store the matrix element a_{ij} in A$(p+q+i-j+1,j)$. Observe that the diagonals of the matrix appear in the rows of the array while the matrix columns correspond to the array columns. The row dimension of the array A must clearly satisfy $adim \geq 2p + q + 1$.

Band Cholesky

The subroutines described in §3.3 for full positive definite symmetric systems also have banded analogs:

Full:	SPOCO(A, adim, n, rcond, z, info)
Banded:	SPBCO(A, adim, n, q, rcond, z, info)

Full:	SPOFA(A, adim, n, info)
Banded:	SPBFA(A, adim, n, q, info)

Full:	SPOSL(A, adim, n, b)
Banded:	SPBSL(A, adim, n, q, b)

Full:	SPODI(A, adim, n, det, job)
Banded:	SPBDI(A, adim, n, q, det)

The arguments to the band routines perform the same function as their counterparts in the full routines. The integer variable q passes the bandwidth. However, the n-by-n matrix A is *not* stored conventionally. For these banded subroutines, the matrix A must be stored as follows:

$$
A \;=\; \begin{bmatrix}
x & x & x & a_{14} & a_{25} & a_{36} & a_{47} & a_{58} \\
x & x & a_{13} & a_{24} & a_{35} & a_{46} & a_{57} & a_{68} \\
x & a_{12} & a_{23} & a_{34} & a_{45} & a_{56} & a_{67} & a_{78} \\
a_{11} & a_{22} & a_{33} & a_{44} & a_{55} & a_{66} & a_{77} & a_{88}
\end{bmatrix}
$$

Here, we are displaying the case $n = 8$, $q = 3$. In this data structure, the matrix element a_{ij} is stored in the array element A(q+1+i-j,j).

Tridiagonal Systems

Consider the tridiagonal system

$$
\begin{bmatrix}
d_1 & e_1 & .. & & 0 \\
c_2 & d_2 & \cdot\cdot & & 0 \\
 & & & & \\
\vdots & \vdots & \ddots & & \vdots \\
 & & & & e_{n-1} \\
0 & 0 & .. & c_n & d_n
\end{bmatrix}
\begin{bmatrix}
x_1 \\ x_2 \\ \\ \vdots \\ \\ x_n
\end{bmatrix}
=
\begin{bmatrix}
b_1 \\ b_2 \\ \\ \vdots \\ \\ b_n
\end{bmatrix}
$$

If the data for this problem is stored in real arrays c, d, e, and b then x can be computed using Gaussian elimination with partial pivoting by invoking the subroutine SGTSL :

$$\text{CALL SGTSL(n, c, d, e, b, info)}$$

If $info = 0$ upon exit, then b is overwritten by the solution x. Otherwise, $info = k > 0$ means that a zero pivot was calculated inhibiting the back-substitution process. The vectors c , d , and e are destroyed in the process.

If the matrix A above is symmetric positive definite with $e(i) = c(i+1)$ for $i = 1,...,n-1$, then x can be determined via tridiagonal Cholesky routine SPTSL:

$$\text{CALL SPTSL(n, d, e, b)}$$

Here, b is overwritten with the solution to $Ax = b$.

Example 3.4-1 (Symmetric Positive Definite Band Solver)

```
      subroutine SPBSOL (A, adim, n,  q, rcond, z, b, info )
      integer  n, q, adim, info
      real A(adim,*), b(*), z(*), rcond

c
c  Solves Ax = b  where A is spd with bandwidth  q.
c
c  On entry:
c      A        real(adim,n)  The nxn symmetric positive
c                             definite matrix  A  stored in
c                             banded form.
c      adim     integer       Row dimension of the array  A.
c      n        integer       Dimension of the matrix  A.
c      b        real(n)       The right-hand side.
c      q        integer       The bandwidth of the matrix A.
c                             Must have q+1 <= adim.
c
c  On exit:
c      A        real(adim,n)  Returns the Cholesky triangle.
c      b        real(n)       Usually returns solution to Ax=b.
c      rcond    real          Returns 1/cond_1( A ) estimate.
c      z        real (n)      Returns unit 1-norm z so
c                             || Az || = rcond|| z || in 1-norm.
c      info     integer       info = 0  =>  x found.
c                             info = -1  =>  A singular, no x.
c
c  Uses LINPACK routines SPBCO and SPBSL.
c  Local Variables:
c      real rcond1
c
      call SPBCO( A , adim , n , rcond , z , info )
      rcond1 = rcond + 1.0e0
      if (info .EQ. 0 .AND. rcond1 .GT. 1.0e0) then
         call SPBSL( A , adim , n , q, b )
      else
         info = -1
      endif
      return
      end
```

Problems for Section 3.4

1. Write a subroutine

        BORDER(n, d, w, alfa, b, info)

that solves the n-by-n symmetric positive definite system $Ax = b$, where A has the form

$$A = \begin{bmatrix} D & w \\ w^T & \alpha \end{bmatrix}$$

Here, $D = diag(d_1 ,..., d_{n-1})$, $w^T = (w_1 ,..., w_{n-1})$, and α is a real scalar. Use info as a trouble indicator.

2. Write a subroutine

        CORNER(n, d, e, alfa, b, info)

that solves the n-by-n symmetric positive definite system $Ax = b$, where A has diagonal $d_1,...,d_n$, sub- and superdiagonal $e_1,...,e_{n-1}$, and is zero everywhere else except $a_{1,n} = a_{n,1} = \alpha$. Use info as a trouble indicator.

3.5 THE QR FACTORIZATION

The QR factorization of an m-by-n matrix A can be calculated with the subroutine SQRDC:

```
CALL SQRDC( A,adim,m,n,qraux,ipvt,work,job )
```

In particular, if

$$A = [\, a_1 ,..., a_n \,]$$

is a column partitioning of A then the routine uses Householder matrices to compute the factorization

$$AP = \left[\, a_{ipvt\,(1)}\ a_{ipvt\,(2)}\ \cdots\ a_{ipvt\,(n)} \,\right] = Q \begin{bmatrix} R \\ O \end{bmatrix}$$

where Q (m-by-m) is orthogonal, R (n-by-n) is upper triangular, and P (n-by-n) is a permutation encoded in ipvt. The argument list is as follows:

A	real (adim,n)	The matrix A. Returns R and Q.
adim	integer	Row dimension of array A.
m	integer	Row dimension of the matrix A.
n	integer	Column dimension of the matrix A.
qraux	real(n)	Returns information needed to get Q.
ipvt	integer (n)	Describes pivoting strategy. Returns permutation information.
work	real (n)	Workspace.
job	integer	Indicates pivoting: $0 =$ no, $1 =$ yes.

166

If `job` = 1 on input, then the column-pivoting strategy is defined by `ipvt` :

$$ipvt(k) > 0 \quad \Rightarrow \quad \text{column } k \text{ of } A \text{ is an "initial" column}$$
$$ipvt(k) = 0 \quad \Rightarrow \quad \text{column } k \text{ of } A \text{ is a "free" column}$$
$$ipvt(k) < 0 \quad \Rightarrow \quad \text{column } k \text{ of } A \text{ is a "final" column}$$

Before the Householder reduction begins, the columns of A are reordered so that the initial columns come first and the final columns come last. The usual column-pivoting strategy is then only applied to the free columns.

There are two standard situations. To compute the QR factorization with no column pivoting, invoke SQRDC as follows:

```
CALL SQRDC( A,adim,m,n,qraux,ipvt,work,0 )
```

To compute the usual QR-with-column-pivoting factorization designate every column to be a "free" column :

```
      do 10 i = 1, n
          ipvt(i) = 0
 10       continue
      call SQRDC( A,adim,m,n,qraux,ipvt,work,1 )
```

Upon exit from SQRDC a host of least square-related computations can be performed with the subroutine SQRSL:

```
SQRSL(A,adim,m,k,qraux,b,Qb,QTb,x,rsd,Ax,job,info)
```

Here,

A	real(adim,n)	Contains Q and R, where $AP = QR$ is computed by SQRDC.
adim	integer	Row dimension of the array A.
m	integer	Row dimension of the matrix A.
k	integer	Indicates "how much" of AP is involved in least squares fit. See below.

qraux	real(n)	Contains information on Q provided by SQRDC.
b	real (m)	Returns a real m-vector b.

To interpret the rest of the calling sequence we need additional notation. Let A_k denote the first k columns of AP and let R_k be the leading k-by-k portion of R. Depending upon what is requested,

Qb	real(m)	Returns the vector Qb.
QTb	real(m)	Returns the vector Q^Tb.
x	real(k)	Returns the minimizer of $\|A_kx - b\|_2$ unless R_k is singular.
rsd	real(m)	Returns the minimum residual $b - A_kx$.
Ax	real(m)	Returns the optimum predictor A_kx.
info	integer	Returns zero if R nonsingular and smallest i so $r_{ii} = 0$ otherwise.
job	integer	Indicates what to do.

The integer job is interpreted as a five-digit decimal ABCDE:

$A \neq 0$	$\Rightarrow$	Qb	returns the vector Qb
$BCDE \neq 0$	$\Rightarrow$	QTb	returns the vector Q^Tb
$C \neq 0$	$\Rightarrow$	x	returns the solution to min $\|A_kx - b\|_2$
$D \neq 0$	$\Rightarrow$	rsd	returns the minimum residual $b - A_kx$
$E \neq 0$	$\Rightarrow$	Ax	returns the predictor A_kx

Separate storage for x, Ax, Qb, QTb, and rsd is not necessary in all cases. Some typical examples suffice to illustrate the overwriting possibilities.

Compute everything:

⇔

```
call SQRSL(A,adim,m,k,qraux,b,Qb,QTb,x,rsd,Ax,11111,info)
```

Just overwrite b with Qb in b:

⇔

```
call SQRSL(A,adim,m,k,qraux,b,b,b,b,b,b,10000,info)
```

Just return Qb but leave b alone:

⇔

```
call SQRSL(A,adim,m,k,qraux,b,Qb,b,b,b,b,10000,info)
```

Just overwrite b with Q^Tb :

⇔

```
call SQRSL(A,adim,m,k,qraux,b,b,b,b,b,b,01000,info)
```

Just return Q^Tb but leave b alone:

⇔

```
call SQRSL(A,adim,m,k,qraux,b,b,QTb,b,b,b,01000,info)
```

Just overwrite $b(1:n)$ with x :

⇔

```
call SQRSL(A,adim,m,k,qraux,b,b,b,b,b,b,00100,info)
```

Just return x and Q^Tb but preserve b:

⇔

```
call SQRSL(A,adim,m,k,qraux,b,b,QTb,x,b,b,00100,info)
```

Just return x and overwrite b with Q^Tb:

⇔

```
call SQRSL(A,adim,m,k,qraux,b,b,b,x,b,b,00100,info)
```

Just return x and overwrite b with $b - A_k x$:

⇔

```
call SQRSL (A, adim, m, k, qraux, b, b, b, x, b, b, 00110, info)
```

Just return x and overwrite b with $A_k x$:

⇔

```
call SQRSL (A, adim, m, k, qraux, b, b, b, x, b, b, 00101, info)
```

Just overwrite b with $b - A_k x$:

⇔

```
call SQRSL (A, adim, m, k, qraux, b, b, b, b, b, b, 00010, info)
```

Just overwrite b with $A_k x$:

⇔

```
call SQRSL (A, adim, m, k, qraux, b, b, b, b, b, b, 00001, info)
```

Example 3.5-1 (Full Rank Least Squares Solver)

```
      subroutine LS (A, adim, m, n, b, x, rcond, z, info, qraux)
      integer adim, m, n, info
      real A(adim,*), b(*), x(*), rcond, z(*), qraux(*)

c
c  Solves  min || Ax - b ||₂ via the factorization  A = QR.
c  A  must not be numerically rank deficient.
c
c  On entry :
c       A       real(adim,n)  The m-by-n matrix A.
c       adim    integer       Row dimension of the array A.
c       m       integer       Row dimension of the matrix A.
c       n       integer       Column dimension of matrix  A.
c       b       real (m)      The real m-vector b.
c
```

```
c On exit :
c      A       real (adim,n)  Returns Q and R.
c      b       real (m)       Returns b - Ax_LS.
c      x       real(n)        Returns the minimizer x_LS of
c                             || Ax - b ||_2.
c      rcond   real           Returns 1/cond_1( A ) estimate.
c      z       real (n)       Returns unit 1-norm z so
c                             || Az || = rcond|| z || in 1-norm.
c      info    integer        info = 0  ⇒ solution found.
c                             info ≠ 0  ⇒ A  rank deficient.
c      qraux   real(n)        returns information on Q.
c
c  Uses LINPACK routines SQRDC, STRCO, and SQRSL
c
c  Local variables
       real rcond1
c
       call SQRDC( A, adim, m, n, qraux, 1, z, 0 )
       call STRCO( A, adim, n, rcond, z, 1 )
       rcond1 = rcond + 1.0e0
       if ( rcond1 .GT. 1.0e0 )   then
           call SQRSL( A,adim,m,n,qraux,b,z,b,x,b,z,110,info )
       else
             info = -1
       endif
       return
       end
```

Example 3.5-2 (Underdetermined System Solver)

```
       subroutine UND(A,adim,m,n,b,x,rcond,z,info,qraux,ipvt)
       integer adim, m, n, info, ipvt(*)
       real A(adim,*), b(*), x(*), z(*), qraux(*), rcond
c
c  Solves  the underdetermined system Ax = b via
c  the factorization AP = QR provided  A  is  not
c  numerically rank deficient.
c
```

```
c   On entry :
c       A      real (adim,n)  The m-by-n matrix A.
c       adim   integer        Row dimension of array  A.
c       m      integer        Row dimension of matrix  A.
c       n      integer        Column dimension of matrix A
c                             must have m < n.
c       b      real (m)       Contains a real m-vector b.
c
c   On exit:
c       A      real(adim,n)   Returns Q and R.
c       b      real(m)        Is destroyed.
c       x      real(n)        Returns the solution to Ax = b.
c       rcond  real           Returns 1/cond₁( A ) estimate.
c       z      real(n)        Returns unit 1-norm z so
c                             || Az || = rcond|| z || in 1-norm.
c       info   integer        info = 0 ⇒ solution found.
c                             info > 0 ⇒ A  rank deficient.
c       qraux  real(n)        Returns information on Q.
c       ipvt   integer(n)     Returns information on P.
c
c   Uses LINPACK routines SQRDC, STRCO, and SQRSL
c
c   Local variables
        real rcond1
        integer k
c
        call SQRDC( A, adim, m , n, qraux, ipvt, z, 1 )
        call STRCO( A, adim, m, rcond, z, 1 )
        rcond1 = rcond + 1.0e0
        if (rcond1 .GT. 1.0e0 )    then
            call SQRSL(A,adim,m,m,qraux,b,z,b,b,b,z,100,info)
            do 30 k = 1 , n
                x(k) = 0.
                if ( ipvt(k) .LE. m ) x(k) = b(ipvt(k))
30          continue
        else
            info = -1
        endif
        return
        end
```

Example 3.5-2 illustrates an important point when using SQRDC with column pivoting. Namely, the user may be obliged to unscramble results as in the 30-loop. This is in contrast to the LINPACK solve routines such as SGESL and SSISL that hide the underlying permutation activity from the user.

Problems for Section 3.5

1. Let A (n-by-n) be given and consider the following factorizations:

(1) $A = QR$	Q orthogonal,	R	upper triangular
(2) $A = QL$	Q orthogonal,	L	lower triangular
(3) $A = RQ$	Q orthogonal,	R	upper triangular
(4) $A = LQ$	Q orthogonal,	L	lower triangular

Let E be the exchange matrix, i.e., the identity with the column ordering reversed. By working with transposes and the exchange matrix it is possible to relate the factorizations (2), (3), and (4) to (1). For example, if $EAE = QR$ is the QR factorization of EAE, then $A = (EQE)(ERE)$ is the "QL" factorization of A. Write a subroutine

```
QR( A, adim, n, Q, qdim, job, work, qraux )
```

that computes the factorization (k) if $job = k$. The orthogonal matrix Q should be returned in Q and the upper (lower) triangular portion of A should be overwritten with R (L). Try to employ the smallest possible workspace work.

2. Let A (m-by-n) and B (m-by-p) be given and assume that $\text{rank}(A) = n$. Write a subroutine

```
PROJ( A, adim, m, n, B, bdim, p, work, qraux )
```

that overwrites B with the matrix $Q_1 Q_1^T B$, where $A = QR$ is the QR factorization of A and Q_1 is comprised of the first n columns of Q.

3. Consider the system

$$\begin{bmatrix} D & E \\ E^T & O \end{bmatrix} \begin{bmatrix} f \\ r \end{bmatrix} = \begin{bmatrix} 0 \\ b \end{bmatrix}$$

where D is n-by-n symmetric positive definite, E is n-by-p with rank p, and b is p-by-1. This can be solved by the following algorithm:

a) Compute the QR factorization

$$E = QR = \begin{bmatrix} Q_1 & Q_2 \end{bmatrix} \begin{bmatrix} R_1 \\ O \end{bmatrix}$$

b) Solve $R_1^T f_1 = b$.

c) Compute

$$Q^T D Q = \begin{bmatrix} D_{11} & D_{12} \\ D_{21} & D_{22} \end{bmatrix}$$

d) Solve $D_{22} f_2 = -D_{21} f_1$ and $R_1 r = -D_{12} f_2 - D_{11} f_1$

e) $f = Q_1 f_1 + Q_2 f_2$

Verify that this algorithm works and implement it in a subroutine of the form

        S220(D, ddim, n, E, edim, p, f, b, work)

D and E need not be preserved. Return r in b. Exploit symmetry as much as possible.

3.6 THE SINGULAR VALUE DECOMPOSITION

The singular value decomposition $A = USV^T$ of an m-by-n matrix A can be computed using the subroutine SSVDC:

```
    call SSVDC( A, adim, m, n, s, e,
   &              U, udim, V, vdim, work, job, info )
```

Here,

A	real (adim, n)	The m-by-n matrix A.
adim	integer	Row dimension of the array A.
m	integer	Row dimension of the matrix A.
n	integer	Column dimension of matrix A.
s	real (n)	Returns the singular values in descending order.
e	real (n)	Work vector.
U	real (udim, m)	Returns the orthogonal matrix U if requested.
udim	integer	Row dimension of the array U.
V	real (vdim, n)	Returns the orthogonal matrix V if requested.
vdim	integer	Row dimension of the array V.
work	real(n)	Work vector.
job	integer	Indicates "how much" of SVD to compute.
info	integer(n)	$info = 0$ implies SVD is found. $info > 0$ implies SVD is not completed.

The variable `job` is a two-digit integer AB that prescribes "how much" of the SVD to compute. For example, U and/or V may not be required. Provision can also be made to form just the first n columns of U, i.e., the matrix U_1 in

$$U = \begin{bmatrix} U_1 & U_2 \end{bmatrix}$$
$$\quad\quad \text{n} \quad \text{m-n}$$

Anything less than the full SVD allows for substantial overwriting by identifying the arrays `U` and/or `V` with the array `A` in the calling sequence. Here are the possibilities:

job	Identification A	U	V	Effect
00	A	A	A	No U or V.
00	A	U	U	No U or V. A saved.
01	A	A	V	Matrix V returned in V.
01	A	A	A	Matrix V returned in top of A.
10	A	U	A	Matrix U returned in U. A saved.
11	A	U	V	Matrices U and V returned in U and V. A saved.
20	A	U	A	Matrix U1 returned in U. A saved.
20	A	A	A	Matrix U1 returned in A.
21	A	U	V	Matrices U1 and V returned in U and V. A saved.
21	A	U	A	Matrices U1 and V returned in U and A. A saved.
21	A	A	V	Matrices U1 and V returned in A and V.

Any option that returns the matrix V in array `A` requires $m \geq n$.

A leading application of the SVD is the linear least squares problem, especially when the data matrix is numerically rank deficient.

Example 3.6-1 (Rank Deficient Least Squares)

```
        subroutine LSSVD( A,adim,m,n,s,e,V,vdim,b,tol x,info )
c
        integer adim, m, n, vdim, info
        real A(adim,*),s(*),e(*),V(vdim,*),b(*),tol,x(*)
c
c   Minimizes || A_r x - b ||_2  where  r is the largest integer
c   such that σ_r(A) ≥ tol  and  A_r is the closest rank r
```

```
c   matrix to A.
c
c   On entry:
c      A       real (adim,n)  The m-by-n matrix  A.
c      adim    integer        Row dimension of the array A.
c                             Must have adim ≥ m.
c      m       integer        Row dimension of the matrix A.
c      n       integer        Column dimension of matrix A.
c      e       real (n)       Work vector.
c      vdim    integer        Row dimension of the array V.
c      b       real(m)        The right-hand side vector.
c      tol     real           A positive tolerance.
c                             tol  ≥ machine precision.
c
c   On exit:
c      A       real (adim,n)  A is destroyed.
c      s       real (n)       Returns the singular values
c                             in descending  order.
c      V       real(vdim,n)   Returns the orthogonal V.
c      x       real(n)        Returns minimizer of
c                             || A_r x - b ||_2.
c      info    integer        info > 0 implies  SVD trouble.
c                             info < 0 implies rank(A) = -info.
c
c   Uses LINPACK routines SDOT, SAXPY, SSVDC.
c
c   Local Variables:
c
        integer i, j
        real ujtb
        call SSVDC( A,adim,m,n,s,e,A,  adim,V,vdim,x,21,info )
c
        if (info .NE. 0 )   then
c
            do 5  i = 1, n
                x(i) = 0.
    5       continue
c
```

```
      do 50 j = 1,n
         if ( s(j) .GE. tol ) then
            ujtb = SDOT( m, a(1,j), 1, b(1), 1)
            call SAXPY( n,ujtb/s(j),V(1,j),1,x(1),1 )
            info = info - 1
         endif
50       continue
      endif
      return
      end
```

Problems for Section 3.6

1. Write a subroutine

```
      POLAR( A, adim, n, U, udim, P, pdim )
```

that computes an orthogonal U and a symmetric nonnegative definite P such that $A = UP$.

2. Let A $(m$-by-$n)$ and B $(p$-by-$n)$ be given and assume that we have $p \leq n \leq m$. Write a Fortran subroutine

```
      SVPART(H,hdim,m,n,p,U1,u1dim,U2,u2dim,V,vdim,s,work)
```

that computes the decomposition

$$
\begin{bmatrix} U_1 & 0 \\ 0 & U_2 \end{bmatrix}^T \begin{bmatrix} B \\ A \end{bmatrix} V = \begin{bmatrix} \Sigma_1 & 0 \\ W & \Sigma_2 \end{bmatrix}
$$

where U_1 $(p$-by-$p)$, U_2 $(m$-by-$m)$, and V $(n$-by-$n)$ are orthogonal, $\Sigma_1 = diag(\sigma_1,...,\sigma_p)$ is n-by-n, and $\Sigma_2 = diag(\sigma_{p+1},...,\sigma_n)$ is $(m-n)$-by-n. Assume that A and B are passed to SVPART in an array H with

$$H = \begin{bmatrix} B \\ A \end{bmatrix}$$

Return the singular values $\sigma_1, ..., \sigma_n$ in the array s .

3.7 DOUBLE PRECISION AND COMPLEX VERSIONS

In the previous sections we described the real single precision LINPACK subroutines. These routines have double precision and complex implementations. The same prefix scheme used by the BLAS applies here:

S   Real Single Precision
D   Real Double Precision
C   Complex Single Precision
Z   Complex Double Precision

Triangular

(*Solve*)

```
STRSL( SA, adim, n, sb, job, info )
DTRSL( DA, adim, n, db, job, info )
CTRSL( CA, adim, n, cb, job, info )
ZTRSL( ZA, adim, n, zb, job, info )
```

(*Condition*)

```
STRCO( SA, adim, n, srcond, sz, job )
DTRCO( DA, adim, n, drcond, dz, job )
CTRCO( CA, adim, n, srcond, cz, job )
ZTRCO( ZA, adim, n, drcond, zz, job )
```

General

(*Factor and Condition*)

```
SGECO( SA, adim, n, ipvt, srcond, sz )
DGECO( DA, adim, n, ipvt, drcond, dz )
CGECO( CA, adim, n, ipvt, srcond, cz )
ZGECO( ZA, adim, n, ipvt, drcond, zz )
```

(Factor)

```
SGEFA( SA, adim, n, ipvt, info )
DGEFA( DA, adim, n, ipvt, info )
CGEFA( CA, adim, n, ipvt, info )
ZGEFA( ZA, adim, n, ipvt, info )
```

(Solve)

```
SGESL( SA, adim, n, ipvt, sb, job )
DGESL( DA, adim, n, ipvt, db, job )
CGESL( CA, adim, n, ipvt, cb, job )
ZGESL( ZA, adim, n, ipvt, zb, job )
```

(Inverse and Determinant)

```
SGEDI( SA, adim, n, ipvt, sdet, swork, job )
DGEDI( DA, adim, n, ipvt, ddet, dwork, job )
CGEDI( CA, adim, n, ipvt, cdet, cwork, job )
ZGEDI( ZA, adim, n, ipvt, zdet, zwork, job )
```

General Banded

(Factor and Condition)

```
SGBCO( SA, adim, n, p, q, ipvt, srcond, sz )
DGBCO( DA, adim, n, p, q, ipvt, drcond, dz )
CGBCO( CA, adim, n, p, q, ipvt, srcond, cz )
ZGBCO( ZA, adim, n, p, q, ipvt, drcond, zz )
```

(Factor)

```
SGBFA( SA, adim, n, p, q, ipvt, info )
DGBFA( DA, adim, n, p, q, ipvt, info )
CGBFA( CA, adim, n, p, q, ipvt, info )
ZGBFA( ZA, adim, n, p, q, ipvt, info )
```

(Solve)

```
SGBSL( SA, adim, n, p, q, ipvt, sb, job )
DGBSL( DA, adim, n, p, q, ipvt, db, job )
CGBSL( CA, adim, n, p, q, ipvt, cb, job )
ZGBSL( ZA, adim, n, p, q, ipvt, zb, job )
```

(Inverse and Determinant)

```
SGBDI( SA, adim, n, p, q, ipvt, sdet, swork, job )
DGBDI( DA, adim, n, p, q, ipvt, ddet, dwork, job )
CGBDI( CA, adim, n, p, q, ipvt, cdet, cwork, job )
ZGBDI( ZA, adim, n, p, q, ipvt, zdet, zwork, job )
```

General Tridiagonal

```
SGTSL( n, sc, sd, se, sb, info )
DGTSL( n, dc, dd, de, db, info )
CGTSL( n, cc, cd, ce, cb, info )
ZGTSL( n, zc, zd, ze, zb, info )
```

Hermitian Positive Definite

(Factor and Condition)

```
SPOCO( SA, adim, n, srcond, sz, info )
DPOCO( DA, adim, n, drcond, dz, info )
CPOCO( CA, adim, n, srcond, cz, info )
ZPOCO( ZA, adim, n, drcond, zz, info )
```

(Factor)

```
SPOFA( SA, adim, n, info )
DPOFA( DA, adim, n, info )
CPOFA( CA, adim, n, info )
ZPOFA( ZA, adim, n, info )
```

(*Solve*)

```
SPOSL( SA, adim, n, sb )
DPOSL( DA, adim, n, db )
CPOSL( CA, adim, n, cb )
ZPOSL( ZA, adim, n, zb )
```

(*Inverse and Determinant*)

```
SPODI( SA, adim, n, sdet, job )
DPODI( DA, adim, n, ddet, job )
CPODI( CA, adim, n, sdet, job )
ZPODI( ZA, adim, n, ddet, job )
```

Banded Hermitian Positive Definite

(*Factor and Condition*)

```
SPBCO( SA, adim, n, q, srcond, sz, info )
DPBCO( DA, adim, n, q, drcond, dz, info )
CPBCO( CA, adim, n, q, srcond, cz, info )
ZPBCO( ZA, adim, n, q, drcond, zz, info )
```

(*Factor*)

```
SPBFA( SA, adim, n, q, info )
DPBFA( DA, adim, n, q, info )
CPBFA( CA, adim, n, q, info )
ZPBFA( ZA, adim, n, q, info )
```

(*Solve*)

```
SPBSL( SA, adim, n, q, sb )
DPBSL( DA, adim, n, q, db )
CPBSL( CA, adim, n, q, cb )
ZPBSL( ZA, adim, n, q, zb )
```

(Inverse and Determinant)

```
SPBDI( SA, adim, n, q, sdet, job )
DPBDI( DA, adim, n, q, ddet, job )
CPBDI( CA, adim, n, q, sdet, job )
ZPBDI( ZA, adim, n, q, ddet, job )
```

Hermitian Positive Definite Tridiagonal

```
SPTSL( n, sd, se, sb, info )
DPTSL( n, dd, de, db, info )
CPTSL( n, sd, ce, cb, info )
ZPTSL( n, dd, ze, zb, info )
```

Hermitian Indefinite

(Factor and Condition)

```
SSICO( SA, adim, n, ipvt, srcond, sz )
DSICO( DA, adim, n, ipvt, drcond, dz )
CSICO( CA, adim, n, ipvt, srcond, cz )
ZSICO( ZA, adim, n, ipvt, drcond, zz )
```

(Factor)

```
SSIFA( SA, adim, n, ipvt, info )
DSIFA( DA, adim, n, ipvt, info )
CSIFA( CA, adim, n, ipvt, info )
ZSIFA( ZA, adim, n, ipvt, info )
```

(Solve)

```
SSISL( SA, adim, n, ipvt, sb )
DSISL( DA, adim, n, ipvt, db )
CSISL( CA, adim, n, ipvt, cb )
ZSISL( ZA, adim, n, ipvt, zb )
```

(Inverse and Determinant)

```
SSIDI(SA, adim, n, ipvt, sdet, inert, swork, job)
DSIDI(DA, adim, n, ipvt, ddet, inert, dwork, job)
CSIDI(CA, adim, n, ipvt, sdet, inert, cwork, job)
ZSIDI(ZA, adim, n, ipvt, ddet, inert, zwork, job)
```

QR Factorization

(Factor)

```
SQRDC(SA, adim, m, n, sqraux, ipvt, swork, job)
DQRDC(DA, adim, m, n, dqraux, ipvt, dwork, job)
CQRDC(CA, adim, m, n, cqraux, ipvt, cwork, job)
ZQRDC(ZA, adim, m, n, zqraux, ipvt, zwork, job)
```

(Solve)

```
SQRSL(SA,adim,m,n,sqraux,sb,sQb,sQTb,sx,srsd,sAx,job,info)
DQRSL(DA adim,m n,dqraux,db,dQb,dQTb,dx,drsd,dAx,job,info)
CQRSL(CA,adim,m,n,cqraux,cb,cQb,cQTb,cx,crsd,cAx,job,info)
ZQRSL(ZA,adim,m,n,zqraux,zb,zQb,zQTb,zx,zrsd,zAx,job,info)
```

Singular Value Decomposition

```
SSVDC(SA,adim,m,n,ss,se,SU,udim,SV,vdim,swork,job,info)
DSVDC(DA,adim,m,n,ds,de,DU,udim,DV,vdim,dwork,job,info)
CSVDC(CA,adim,m,n,ss,se,CU,udim,CV,vdim,cwork,job,info)
ZSVDC(ZA,adim,m,n,ds,de,ZU,udim,ZV,vdim,zwork,job,info)
```

Problems for Section 3.7

1. Some complex problems can be cast in real form at the expense of increased dimension. For example, the equation

$$(A_1 + iA_2)(x_1 + ix_2) = (b_1 + ib_2)$$

is equivalent to the real system

$$\begin{bmatrix} A_1 & -A_2 \\ A_2 & A_1 \end{bmatrix} \begin{bmatrix} x_1 \\ x_2 \end{bmatrix} = \begin{bmatrix} b_1 \\ b_2 \end{bmatrix}$$

The complex least squares problem

$$min \; || \; (A_1 + iA_2)(x_1 + ix_2) \; - \; (b_1 + ib_2) \; ||_2$$

is equivalent to the real arithmetic least squares problem

$$min \; \left\| \begin{bmatrix} A_1 & -A_2 \\ A_2 & A_1 \end{bmatrix} \begin{bmatrix} x_1 \\ x_2 \end{bmatrix} - \begin{bmatrix} b_1 \\ b_2 \end{bmatrix} \right\|_2$$

Write real arithmetic subroutines CGECO, CPOCO, CSICO, and CQRDC that compute the real and imaginary parts of the following factorizations:

CGECO:	$A = LU,$	A is a general square complex matrix
CPOCO:	$A = LL^H,$	A is Hermitian positive definite and complex
CSICO:	$A = LDL^H,$	A is Hermitian, indefinite, and complex
CQRDC:	$A = QR,$	A is a general rectangular complex matrix

2. Write the double precision and complex single precision versions of the subroutine SGSOL in Example 3.2-1.

3. Write the double precision and complex single precision versions of the subroutine LS in Example 3.5-1.

MATLAB

MATLAB is an interactive system in which it is possible to express matrix algorithms at a very high level. For example, QR factorizations and eigenvalue decompositions are "one-liners." Submatrices and block matrices are easily manipulated.

Our aim is to convey a sense of the MATLAB language. We do not cover MATLAB graphics or those aspects of MATLAB that involve interaction with the underlying operating system. These matters are best left to *The MATLAB User's Guide* [7] and the excellent on-line help facility.

4.1 BASICS

Once a MATLAB session begins, the user can create matrices and vectors and perform computations that involve them. We describe some of the ways that matrices and vectors can be set up, some elementary operations that can be performed, and how to display results.

Setting Up Matrices

Matrices can be explicitly introduced with assignment statements of the form

```
A = [ 1 1 1 1 ; 1 2 3 4 ; 1 3 6 10 ; 1 4 10 20 ] ;
```

This instruction assigns the 4-by-4 Pascal matrix to A:

$$
A \;=\; \begin{bmatrix} 1 & 1 & 1 & 1 \\ 1 & 2 & 3 & 4 \\ 1 & 3 & 6 & 10 \\ 1 & 4 & 10 & 20 \end{bmatrix}
$$

In general, to set up an m-by-n matrix A execute the statement

```
A = [{ row 1} ; { row 2} ; . . . ; { row m}  ] ;
```

where the n entries for each row are separated by one or more blanks. It is also legal to make input "look like" output when setting up matrices.

```
A = [ 1   2
      3   4 ]
```

is the same as A = [1 2; 3 4] .

Simple Output

If a statement ends with a semicolon, then the result of the statement is *not* displayed. Thus, if you enter

```
A = [1 2 ; 3 4 ]
```

then the system responds with

```
A =
   1   2
   3   4
```

Unless suppressed by a semicolon, any change in a matrix prompts the printing of the entire matrix. Thus, if you now enter

```
A(1,2)   =   A(2,1)
```

then

```
A =
   1   3
   3   4
```

is displayed. Just typing

```
A
```

displays *A*.

Types and Dimension

The complex array is the only data type in MATLAB. Special cases are acknowledged in the displaying of results. Real arrays are displayed as real. Integer arrays are displayed as integers.

Dimensioning is handled automatically in MATLAB. Suppose you set

```
B = [ 1 2 3 ; 4 5 6 ];
```

and then subsequently overwrite this matrix with the 2-by-2 identity:

```
B = [ 1 0 ; 0 1 ];
```

The system "knows enough" to recognize that B has changed from a 2-by-3 matrix to a 2-by-2 matrix.

There are other important aspects associated with MATLAB's automatic dimensioning capability. If C is uninitialized then the command

```
C(3,2) = 2
```

prompts the response

```
C =
    0  0
    0  0
    0  2
```

In other words, MATLAB makes C "just large enough" so that the assignment makes sense. If you now enter

```
C(1,3) = 0
```

then

```
C =
    0  0  0
    0  0  0
    0  2  0
```

Subscripts

Subscripts in MATLAB must be positive. Fractional subscripts are "floored":

$$A(1.2, 3.5) = 10 \quad \Leftrightarrow \quad A(1,3) = 10$$

Vectors and Scalars

Vectors and scalars are "special" matrices and thus can be set up using matrix assignment. The commands,

```
x = [ 1; 2; 3 ];
```

or

```
x = [ 1 2 3 ]';
```

establish x as a column vector while

```
x = [ 1 2 3 ];
```

creates a row vector. Scalars are 1-by-1 matrices, so the statement

```
c = [ 3 ];
```

is a scalar assignment. However, in the case of scalars the brackets can be omitted:

```
c = 3 ;
```

Size and Length

The "current" row and column dimensions of an array can be computed with the built-in function **size**:

```
[ m,n ] = size(A)
```

This assigns the row dimension of A to m and the column dimension to n. As we shall see, many of MATLAB's built-in functions return multiple values.

If v is either a row or column vector then

```
n = length(v)
```

returns the dimension of *v*.

Continuation

Sometimes it is not possible to put an entire MATLAB instruction on a single line. When this is the case, two adjacent periods can be used to signal continuation. Thus

```
A = [ 1 2 3 ; 4 5 6 ;  ..
      7 8 9 ];
```

is the same as

```
A = [ 1 2 3; 4 5 6; 7 8 9 ];
```

When spreading an instruction over several lines, break the instruction in a "natural" place.

Variable Names

Variable names in MATLAB must consist of nineteen or fewer characters. The name must begin with a letter. Any letter, number or underscore may follow, e.g., sym_part1. Upper and lower cases *are* distinguished when a MATLAB session begins. However, the case sensitivity status can be toggled by issuing the command **casesen**. Thus if upper and lower cases are currently distinguished then

```
casesen
A = [ 1 2; 3 4];
a(2,2) = 1
```

produces

```
A =
    1    2
    3    1
```

Addition, Multiplication, Exponentiation

If A and B are matrices then

$$C = A + B$$

sets C to be the sum while

$$C = A*B$$

is the product. Of course, the dimensions of A and B must be appropriate for these operations. Otherwise an error message is displayed.

To raise a square matrix A to a power r enter

$$C = A^{\wedge}r$$

This operation works for *any* scalar r. In particular

$$C = A^{\wedge}(-1)$$

assigns the inverse of A to C.

If t is a scalar and A is a matrix then scalar-matrix multiplication is effected by

$$C = t*A$$

Matrix/vector computations are equally easy to express. Consider the computation of the vector

$$f = c_1 Ab + c_2 A^2 b + c_3 A^3 b$$

where A is a matrix, b is a vector, and the c_i are scalars. Here are two equivalent ways to compute f :

```
f = A*(A*(c(3)*A*b + c(2)*b ) + c(1)*b)

f = (c(3)*A^3 + c(2)*A^2 + c(1)*A)*b
```

Although these two statements are equivalent, the first turns out to be much more efficient than the second. See the discussion of **flops** later on in this section.

Transpose

The conjugate transpose of a matrix can be obtained using a single quote:

```
B = A'
```

This sets $B = A^T$ if A is real. Thus, if x is an n-vector, then

```
A = x*x'
```

is an n-by-n rank-one matrix. A column vector x can be transformed into a row vector by executing

```
x = x'
```

A common mistake in MATLAB is to confuse row vectors with column vectors and vice versa.

The "Colon Method" for Setting Up Vectors

Colons can be used to prescribe vectors whose components differ by a fixed increment. Here are some examples:

```
v = 1:3        ⇔    v = [ 1 2 3 ]
v = 3:-1:0     ⇔    v = [ 3 2 1 0 ]
v = 1:2:10     ⇔    v = [ 1 3 5 7 9 ]
```

Variables can also be used to specify the limits, e.g.,

```
v = i:j
```

In general, if a unit increment is required then the assignment takes the form

$$v = \{ \text{ start } \} : \{ \text{ stop} \}$$

while

$$v = \{ \text{ start } \} : \{ \text{ increment } \} : \{ \text{ stop } \}$$

should be used for increments other than unity. The starting value, increment, and terminating value may be determined by expressions and they need not be integers.

Remember that the colon method generates *row* vectors.

Logarithmically Spaced Vectors

Just as the colon method can be used to generate "arithmetically regular" vectors, the built-in function **logspace** can be used to generate logarithmically regular vectors. In particular, if

$$w = \text{logspace}(d1, d2, n)$$

then

$$w(i) = 10 \;\hat{}\; [d1 + (i-1)(d2-d1)/(n-1)]$$

for $i = 1{:}n$. Thus,

$$w = \text{logspace}(0,3,4) \quad \Leftrightarrow \quad w = [1 \ 10 \ 100 \ 1000]$$

If **logspace** is invoked with just two arguments, it is assumed that $n = 50$. Note that **logspace** generates *row* vectors.

Generation of Special Matrices

MATLAB has a number of built-in functions that can be used to generate frequently occurring matrices:

$A = \text{eye}(m,n) \quad \Rightarrow \quad A$ *m*-by-*n*, ones on diagonal, zeros elsewhere

$A = \text{zeros}(m,n) \quad \Rightarrow \quad A$ *m*-by-*n*, zeros everywhere

$A = \text{ones}(m,n) \quad \Rightarrow \quad A$ *m*-by-*n*, ones everywhere

For square versions only one argument is needed:

```
A = eye(n)
A = zeros(n)
A = ones(n)
```

If dimension can be inferred from context, then **ones** can be used without any arguments. For example,

```
A = A + ones
```

adds a one to each entry in A regardless of its size.

Matrix arguments can be used to specify dimension information to **eye**, **zeros**, and **ones**. For example, if B is an m-by-n matrix then

```
A = ones(B)
```

assigns the m-by-n matrix of ones to A. (B must not be a 1-by-1 matrix for this to work.) Similar comments apply to

```
A = eye(B)
```
and
```
A = zeros(B)
```

Toeplitz matrices can be generated using **toeplitz**. If r and c are n-vectors and

```
A = toeplitz(c,r)
```

then $a_{ij} = c_{i-j+1}$ if $i \geq j$ and $a_{ij} = r_{i-j+1}$ if $j > i$. To generate a Hermitian Toeplitz matrix make sure c_1 is real and use

```
A = toeplitz(c)
```

for then $a_{ij} = c_{i-j+1}$ if $i \geq j$ and $a_{ij} = conj(c_{i-j+1})$ if $j > i$.

Random Matrices

It is possible to generate matrices and vectors of random numbers:

```
A = rand(m,n)    ⇒ A m-by-n, random elements
A = rand(n)      ⇒ A m-by-n,  random elements
A = rand(B)      ⇒ A same size as B, random elements
```

There are two probability distributions from which random numbers can be selected: the uniform distribution on [0,1] and the normal distribution with mean 0 and variance 1 . The commands

```
                rand('uniform')
```
and
```
                rand('normal')
```

are used to establish the "current" distribution. The uniform distribution is in force when a MATLAB session begins.

To clarify these notions, the commands

```
        rand ('normal')
        A = rand (3,3) ;
        rand ('uniform')
        b = rand(3,1);
```

generate a 3-by-3 matrix A with normally distributed entries and a 3-by-1 vector b with uniformly distributed entries.

It is possible to manipulate the five-digit integer seed that is used by the random number generator.

```
        n = rand('seed')    ⇒      n = current seed
        rand('seed',n)      ⇒      current seed = n
```

This is useful when "random" experiments have to be repeated. The seed is zero when a MATLAB session begins.

Complex Matrices

If A and B are real matrices with the same dimension and $i^2 = -1$, then the complex matrix $C = A + iB$ can be generated as follows:

```
C = A + sqrt(-1)*B ;
```

If several operations involving $sqrt(-1)$ are required, it is handy to set

```
i = sqrt(-1);
```

for then assignments of the form

```
C = A + i*B
```

assume the appearance of ordinary mathematics. For example,

```
V = [ 1+i ; 1-i  ]
```

establishes v to be a complex column vector. When an expression is used to define an array entry, it is important not to have blanks within the expression. The command

```
V = [ 1 + i ; 1 - i ]
```

generates a 2-by-2 matrix:

$$V = \begin{bmatrix} 1 & i \\ 1 & -i \end{bmatrix}$$

Pointwise Operations

Element operations between two matrices of the same size are also possible:

```
C = A.*B        ⇒        c_ij = a_ij b_ij
C = A.\B        ⇒        c_ij = b_ij / a_ij
```

$$C = A.*B \quad \Rightarrow \quad c_{ij} = a_{ij} b_{ij}$$
$$C = A.\backslash B \quad \Rightarrow \quad c_{ij} = b_{ij} / a_{ij}$$

$$C = A./B \quad \Rightarrow \quad c_{ij} = a_{ij}/b_{ij}$$
$$C = A.^\wedge B \quad \Rightarrow \quad c_{ij} = a_{ij}^{\ b_{ij}}$$

In addition, if e is a scalar and A is a matrix then

$$C = A.^\wedge e \quad \Rightarrow \quad c_{ij} = a_{ij}^{\ e}$$

Of course, the above statements make sense only if the corresponding point operations make sense.

Another example of the "dot" operator is

$$C = A.'$$

which sets C to be the unconjugated tranpose of A, i.e., $c_{ij} = a_{ji}$.

More on Input/Output

The format for printed output can be controlled using the following commands:

format short	$\Rightarrow$	5-digit fixed point style
format short e	$\Rightarrow$	5-digit floating point style
format long	$\Rightarrow$	15-digit fixed point style
format long e	$\Rightarrow$	15-digit floating point style
format hex	$\Rightarrow$	hexadecimal

Once a format is specified in a MATLAB session, it remains in force until a new format command is issued. When a MATLAB session begins, the **short** format prevails.

Additional output techniques are detailed in §4.7.

Machine Precision

At the start of a MATLAB session, the variable **eps** contains the effective machine precision, i.e., the smallest floating point number such that if

```
x = 1 + eps
```

then $x > 1$. It is handy to use **eps** for zero checking in the presence of roundoff. It is legal to alter the value of **eps** but it is not recommended.

Flops

The variable **flops** maintains the current number of flops that have been performed since the MATLAB session began. The command

```
flops(0)
```

resets this counter to zero. For real operands, each arithmetic operation counts as a single flop. For complex operands, addition and subtraction count as two flops while multiplications and divisions each count as six flops.

To illustrate the flop notion, suppose A (10-by-10), b (10-by-1), and c (1-by-1) are initialized and real. The following table lists the number of flops required for certain basic computations:

Operation	Flops
c*A	100
A*b	200
A*A	2000
b*b'	200
A*b*b'	400
b*b'*A	2200
b*(b'*A)	400

When confronted with two operations of equal precedence, MATLAB evaluates from left to right. Note from the last two entries in the table that it is sometimes appropriate to insert parentheses to ensure efficient evaluation.

Problems for Section 4.1

1. Give a one-liner for setting up

$$v = [\,100.5 \quad 90.5 \cdots 10.5 \ .5\,]$$

2. Suppose n is a positive integer. Give a one-liner that generates the vector

$$v = [\,2 \ 4 \ 8 \ 16 \cdots 2^n\,]$$

3. Suppose the statements

```
A = rand(10)
x = rand(10,1)
```

have been executed. How many flops are required to execute each of the following statements:

```
y = A*A*x
y = (A*A)*x
y = A*(A*x)
B = zeros(10,10)*A
```

4. Suppose n is a positive integer and *low* and *high* are real numbers that satisfy *low* < *high*. Write a one-liner that assigns to A an n-by-n matrix whose entries are chosen from the uniform distribution on the interval [*low*, *high*].

5. Assume that n (integer), w (complex), and x (n-by-1) are given. For each of the following examples write a one-liner that generates the n-by-n matrix $A = (a_{ij})$:

(a) $a_{ij} = w^{(i-1)(j-1)}$
(b) $a_{ij} = 1/(i+j-1)$
(c) $a_{ij} = x_i^{\,j-1}$

4.2 LOOPS AND CONDITIONALS

MATLAB has all the usual control structures: **for**-loops, **while**-loops, and **if-then-else**. They work in predictable ways.

For-Loops

In MATLAB, **for**-loops have the form

$$\textbf{for}\ \{\ var\ \} = \{\ row\ vector\ of\ counter\ values\ \}$$

$$\{\ statements\ \}$$

end

For example,

```
for i = 1:3
    x(i) = i;
end
```

is equivalent to

```
x = [ 1;2;3 ];
```

A **for**-loop can also be used to set up matrices. If d is an n-vector and e is an $(n-1)$-vector then the instructions

```
B = zeros(n);
for i = 1:n-1
    B(i,i) = d(i);
    B(i,i+1) = e(i);
end
B(n,n) = d(n)
```

set up an n-by-n upper bidiagonal matrix. Because it is not followed by a semicolon, the last statement prompts the printing of the matrix B. An excessive amount of output would result if the semicolons in the loop body were deleted. Negative increments are also possible in a for-loop. The segment

```
F  =  c(d)*A;
for  k  =  d-1:-1:1
    F  =  A*F  +  c(k)*A  ;
end
```

computes the matrix polynomial

$$F = c_1A + c_2A^2 + \cdots + c_dA^d$$

using Horner's rule.

Loops can be nested. If n has been initialized to some positive integer then

```
for  i  =  1:n
    for  j  =  1:n
        A(i,j)  =  1/(i + j - 1);
    end
end
```

sets A to be the n-by-n Hilbert matrix.

The values that the counter variable takes on are prescribed by the vector of counter values that appears in the **for** statement. This vector need not be specified by the colon notation as in the above illustrations. For example,

```
x  =  [ 0  0  0 ];
w  =  [ 3  1  9  2  6  2  3 ];
v  =  [ 1  3    4    ];
for  i  =  v
    x(i)  =  w(i);
end
```

is equivalent to

```
x  =  [ 3   0   9   2 ];
```

It is also legal for the count variable to take on vector values. If *A* is a matrix then

```
for k = A
    i = k
end
```

prints the columns of *A* one at a time.

Relations

A relation has the form

{ *matrix expression* }   { *relational operator* }   { *matrix expression* }

where the following are legal relational operators:

==	equals
~=	not equals
<	less than
< =	less than or equal to
>	greater than
> =	greater than or equal to

The value of a relation is a 0-1 matrix that has the same size as the matrix expressions in the relation. The 1's appear everywhere the relation holds. Thus

```
A = [ 1 2 ; 3 4 ];
B = [ 1 2 ; 2 4 ];
T =   A == B
```

produces the output

```
T =

    1  1
    0  1
```

We again stress that there is only one variable type in MATLAB: complex matrices. The matrix T is not a Boolean array. It is a complex matrix that happens to be storing 0's and 1's.

"and," "or," "not"

The **and** (&) , **or** (|) , and **not** (~) operations are also possible between 0-1 matrices of equal dimension. Thus, if

```
T1 = [ 1 1 ; 0 1]

T2 = [ 1 0 ; 0 0]
```

then

```
T = T1 & T2
```

prompts

```
T =
    1    0
    0    0
```

while

```
T = T1 | T2
```

sets

```
T =
    1    1
    0    1
```

and

```
T =   ~T1
```

renders

```
T =
   0   0
   1   0
```

It is possible, but not good practice, to apply logical operations to matrices that are not 0-1.

"if" Constructions

The simplest **if** statement has the form

if { *relation* }

{ *statements* }

end

Thus,

```
if print_flag > 0
   A
end
```

prints *A* if *print_flag* is positive.

If-then-else constructs are also possible. Suppose *n* is a positive integer and that we want to set *A* to be the upper triangular matrix that has 1's on the diagonal and -1' s above the diagonal. The following instructions accomplish this:

```
for i = 1:n
   for j = 1:n
      if i < j
         A(i,j) = -1;
      elseif i > j
         A(i,j) = 0;
      else
         A(i,j) = 1;
      end
   end
end
```

If the tested relation compares scalars, as in the above examples, then the **if** "reads" in the familiar way; but a relation can also compare matrices. In this case the tested relation is regarded as true if and only if it is true in each element. For example, if A and B are real matrices then

```
if A > B
   C = A;
end
```

assigns A to C if and only if $a_{ij} > b_{ij}$ for all i and j.

In general, the tested condition in an **if** statement is an expression and as such is just another matrix. If this matrix has no zero elements, then the relation is "true."

The "any" and "all" Functions

The **any** and **all** functions can be used to sense the zero-nonzero structure of a matrix. They are often used in more complicated conditionals that involve matrices.

If **any** is applied to a vector it returns a "1" if any component of the vector is nonzero. Otherwise, it returns a "0." Thus, if v is a vector then

```
any( v > ones )
```

returns "1" if any $v_i > 1$ and it returns "0" otherwise. (Recall that $v > 1$ is a 0-1 vector.)

If the function **all** is applied to a vector then it returns a "1" if all the entries are nonzero. Otherwise it returns a "0." Thus,

```
all( v > ones )
```

returns "1" if each $v_i > 1$ and it returns "0" otherwise.

The functions **any** and **all** can be applied to matrices. The result is a 0-1 vector in which the ith component reflects the status of the ith column. For example, if

$$A = [\ a_1, ..., a_n\ ]$$

and each a_i is an m-vector, then

$$w = \texttt{any(A)} \quad \Rightarrow \quad w = [\, any(\,a_1\,) \,,..., any(\,a_n\,)\,]$$
$$w = \texttt{all(A)} \quad \Rightarrow \quad w = [\, all(\,a_1\,) \,,..., all(\,a_n\,)]$$

While-Loops

While-loops are also possible and have the general form:

while { *relation* }

{ *statements* }

end

Here is a while-loop that is equivalent to the statement

```
w = any(v)
```

assuming that v is a vector:

```
w = 0
k = 1
while ( k <= length(v)  &  w == 0 )
    if v(k)  ~= 0
        w = 1
    end
    k = k+1
end
```

The tested relation in a while-loop need not be scalar-valued. The following MATLAB segment prints a sequence $\{A_i\}$ of random 2-by-2 matrices with the property that $A_{i+1} > A_i$:

```
A = zeros(2);
B = rand(2,2);
while ( B > A )
    A = B
    B = rand(2,2);
end
```

In a while-loop, the body is executed if the tested condition has no nonzero entries.

The "break" Command

The **break** command can be used to terminate a loop. The following MATLAB segment prints all Fibonnaci numbers (see §1.3) less than 1000:

```
fib(1) = 1
fib(2) = 1
for j = 3:1000
    z = fib(j-1) + fib(j-2);
    if z >= 1000
        break
    end
    fib(j) = z
end
```

If a **break** is encountered in the body of a loop, then the loop is immediately terminated.

The use of **break** should be minimized in the interests of readability. It is usually possible to avoid **break** merely by using a while-loop instead of a for-loop. Here is a rewrite of the above Fibonnaci computation:

```
fib(1) = 1
fib(2) = 1
j = 2;
z = fib(1) + fib(2)
while z < 1000
    j = j+1;
    fib(j) = z
    z = fib(j) + fib(j-1)
end
```

Problems for Section 4.2

1. The n-by-n Frank matrix $F_n = (f_{ij})$ is an upper Hessenberg matrix whose nonzero elements are given by $f_{ij} = n - j + 1$ if $j > i$, and by $f_{ij} = n - j$, whenever $j = i - 1$. Write a MATLAB segment that computes F_n, given a positive integer n and test it with $n = 10$.

2. The n-by-n Pascal matrix $P = P_n = (p_{ij})$ is defined by $p_{i,1} = p_{1,i} = 1$ for $i = 1,\ldots,n$ and $p_{ij} = p_{i,j-1} + p_{i-1,j}$ otherwise. Write a MATLAB segment that generates P_n, given n. Use your segment to print the largest Pascal matrix that can be *exactly* represented in floating point on your system.

3. Let $b(1:p)$ be an initialized 0-1 vector and assume A is n-by-n. Write an efficient MATLAB segment that computes $F = A^r$, where

$$r = (b_p \ldots b_2 b_1)_2$$

4. Assume that A is n-by-n and has the property that $\lim A^p = 0$. Write a MATLAB segment that prints the smallest nonnegative integer p so that if $B = A^p$ then

$$-.5 < b_{ij} < 1$$

for all i and j. Use **any** and apply it to $A = [-.2 \ 20 \ ; 0 \ .99]$.

4.3 WORKING WITH SUBMATRICES

The manipulation of submatrices is elegantly handled in MATLAB. In this section we show how to set up block matrices as well as how to alter designated blocks.

Setting Up Block Matrices

Suppose $A11, A12, A21, A22, A31$, and $A32$ are matrices and that their dimensions are such that

$$
A = \begin{bmatrix} A11 & A12 \\ A21 & A22 \\ A31 & A32 \end{bmatrix}
$$

makes sense. This block matrix can be set up as follows:

```
A = [ A11 A12 ; A21 A22 ; A31 A32 ]
```

The general command for setting up a block matrix has the form

```
A = [ { block row 1 }; ... ; { block row p } ]
```

The block rows can be specified by expressions but they each must have the same number of columns. For example, if A (n-by-n) and b (n-by-1) are given, then

```
M = [ A   A*b ; b'*A   b'*A*b   ]
```

sets up the matrix

$$
M = \begin{bmatrix} A & Ab \\ b^H A & b^H Ab \end{bmatrix}
$$

When setting up block matrices it is crucial that no blanks appear within the individual block expressions. For example

```
M = [ A A * b ; b'*A  b'*A*b  ]
```

would result in an error because of the spaces in the (1, 2) block expression.

As another example, here is a loop that sets up a random block diagonal matrix $A = diag(A_1,...,A_k)$, where each A_i is p-by-p.

```
A = rand(p,p)
for i = 2:k
    [ m,m] = size(A);
    A = [ A zeros(m,p); zeros(p,m) rand(p,p) ]
end
```

The Empty Matrix

The *empty matrix* [ ] is often useful for initialization of certain matrix computations. Here is a loop that computes

```
B = A(:,n:-1:1)
```

that sets up a column-reversed version of an n-by-n matrix A :

```
B = [];
for j = n:-1:1
    B = [ B A(:,j) ];
end
```

Designating Submatrices

If A is m-by-n and the integers $i, j, p,$ and q satisfy

$$1 \le i \le j \le m$$
$$1 \le p \le q \le n$$

then

$$B = A(i:j,p:q) \quad \Rightarrow \quad B = \begin{bmatrix} a_{ip} & \cdots & a_{iq} \\ \vdots & & \vdots \\ a_{jp} & \cdots & a_{jq} \end{bmatrix}$$

$$B = A(i , p:q) \quad \Rightarrow \quad B = [a_{ip} ,...., a_{iq}]$$

$$B = A(i:j,p) \quad \Rightarrow \quad B = \begin{bmatrix} a_{ip} \\ \vdots \\ a_{jp} \end{bmatrix}$$

$$B = A(i:j,:) \quad \Rightarrow \quad B = \begin{bmatrix} a_{i1} & \cdot\cdot & a_{in} \\ \vdots & & \vdots \\ a_{j1} & \cdot\cdot & a_{jn} \end{bmatrix}$$

$$B = A(:,p:q) \quad \Rightarrow \quad B = \begin{bmatrix} a_{1p} & \cdot\cdot & a_{1q} \\ \vdots & & \vdots \\ a_{mp} & \cdot\cdot & a_{mq} \end{bmatrix}$$

Of course, expressions can be part of the colon notation as can negative increments:

$$B = A(i:i+1, 3:-1:2) \quad \Rightarrow \quad B = \begin{bmatrix} a_{i3} & a_{i2} \\ a_{i+1,3} & a_{i+1,2} \end{bmatrix}$$

So far, all of our submatrix examples have involved the extraction of contiguous entries from the parent matrix. More generality is possible. If A is a 5-by-4 matrix then

$$B = A([1 \ 3 \ 5] , [2 \ 4]) \quad \Rightarrow \quad B = \begin{bmatrix} a_{12} & a_{14} \\ a_{32} & a_{34} \\ a_{52} & a_{54} \end{bmatrix}$$

It is often handy to use variables to hold the vector of indices that define a submatrix. For example, the above 3-by-2 submatrix can be generated as follows:

```
rows = [ 1 3 5 ];
cols = [ 2 4 ];
B = A(rows,cols);
```

Attempting to access a "nonexistent" matrix entry invokes messages of the form

```
"index exceeds matrix dimensions"
```

```
"index into matrix is negative or zero"
```

Assignment to a Submatrix

It is also possible to make an assignment to a designated submatrix. For example, if A (m-by-n), v (n-by-1), and w (m-by-1) are given and the integers p and q satisfy $1 \leq p \leq m$ and $1 \leq q \leq n$ then

```
A(p,:) = v'
```

replaces the pth row of A with v^H while

```
A(:,q) = w
```

replaces the qth column of A with w.

Here is one way to compute a Jacobi rotation in rows p and q of an n-by-n matrix A, assuming that the cosine-sine pair (c,s) is known:

```
J = eye(A)
J([p q],[p q]) = [ c s; -s c ]
A = J'*A*J
```

This is an $O(n^3)$ computation. The following accomplishes the same thing in $O(n)$ operations:

```
A([ p q ],:) = [ c s ; -s c]'*A([ p q ],:)
A(:,[ p q ]) = A(:,[ p q ])*[ c s ; -s c ]
```

In conjunction with submatrix assignment, the empty matrix is handy for deleting "interior" rows or columns. For example, if A is m-by-n and the indices i, j, p, and q satisfy $1 < i \leq j < m$ and $1 < p \leq q < n$ then

```
A(i:j,:) = []   ⇔   A = [A(1:i-1,:);A(j+1:m,:)]
```

```
A(:,p:q) = []   ⇔   A = [A(:,1:p-1) A(:,q+1:n)]
```

Kronecker Product

The Kronecker product of two matrices A and B can be calculated using the built-in function **kron**:

$$C = \text{kron}(A, B) \quad \Rightarrow \quad C = (a_{ij} B)$$

This is equivalent to

```
C = [];
[m,n] = size(A);
for j = 1:n
    ccol = []
    for i = 1:m
        ccol = [  ccol ; A(i,j)*B ];
    end
    C = [ C ccol ];
end
```

Thus, if

```
C = kron( [ 1 2 ] , [ 1 2 ; 3 4 ] )
```

then

$$C = \begin{bmatrix} 1 & 2 & 2 & 4 \\ 3 & 4 & 6 & 8 \end{bmatrix}$$

Notice how quickly dimensions get large when **kron** is invoked.

Turning Matrices into Vectors and Vice Versa

Finally, if A is m-by-n then

```
x = A(:)
```

is equivalent to

```
x = []
for j = 1:n
    x = [ x ; A(:,j) ]
end
```

The reverse of this matrix-to-vector operation is also possible. If A is a defined m-by-n matrix and v is an mn-by-1 vector, then

```
A(:) = v
```

is the same as

```
for j = 1:n
    A(:,j) = v( (j-1)*m+1:j*m )
end
```

To illustrate these matrix/vector conversions, here is a long-winded way of computing the transpose of a 2-by-2 matrix A:

```
x = A(:);
x( [2 3] ) = x( [3 2] );
A(:) = x;
```

Problems for Section 4.3

1. Write a MATLAB segment that overwrites A (m-by-n) with PA, where P is any permutation chosen so that all the zero rows of PA are at the bottom.

2. Suppose k, a positive integer, and A (n-by-n) are given. Write a MATLAB one-liner that generates the k-by-k block diagonal matrix $C = diag(A, ..., A)$.

3. Write a MATLAB one-liner that is equivalent to the following program segment which replaces every "interior" matrix entry of A (m-by-n) with the average of its four neighbors.

```
for i = 2:m-1
    for j = 2:n-1
        B(i,j) = (A(i-1,j) + A(i,j-1) ..
                 + A(i+1,j) + A(i,j+1))/4;
    end
end
A(2:m-1,2:n-1) = B(2:m-1,2:n-1)
```

4. Write a MATLAB segment that overwrites

$$A = \begin{bmatrix} A_{11} & A_{12} \\ A_{21} & A_{22} \end{bmatrix}$$

with

$$A = \begin{bmatrix} A_{22} & -A_{21}^H \\ -A_{12}^H & A_{11} \end{bmatrix}$$

where each A_{ij} is n-by-n.

4.4 BUILT-IN FUNCTIONS

Numerous built-in functions make it possible to couch matrix computations at a fairly high level in MATLAB. Some of the more elementary functions are discussed in this section, whereas those that involve matrix factorizations are covered in §4.6.

"abs," "real," "imag," "conj"

If A is m-by-n then

$$
\begin{array}{lll}
B = \text{abs}(A) & \Rightarrow & B = (\,|\,a_{ij}\,|\,) \\
B = \text{real}(A) & \Rightarrow & B = (\,real(\,a_{ij})\,) \\
B = \text{imag}(A) & \Rightarrow & B = (\,imag(a_{ij})\,) \\
B = \text{conj}(A) & \Rightarrow & B = real(A) - i\,imag(A),\ i^2 = -1
\end{array}
$$

Thus,

```
A = conj(A)'
```

overwrites A with its unconjugated transpose.

Before we go any further we remind the reader that the names of MATLAB's *built-in functions are in lower case*. A reference of the form

```
B = ABS(A);
```

would result in an error unless **casesen** is appropriately toggled.

Norms

Various norms are easily computed in MATLAB. If A (m-by-n) is given and real then

$$
\begin{array}{lll}
\texttt{t = norm(A)} & \Rightarrow & t = \|A\|_2 \\
\texttt{t = norm(A,1)} & \Rightarrow & t = \|A\|_1 \\
\texttt{t = norm(A,'INF')} & \Rightarrow & t = \|A\|_\infty \\
\texttt{t = norm(A,'FRO')} & \Rightarrow & t = \|A\|_F
\end{array}
$$

If A is complex then $\texttt{t = norm(A,1)}$ implies

$$t = \| abs(real(A)) + abs(imag(A)) \|_1$$

while $\texttt{norm(A,'INF')}$ implies

$$t = \| abs(real(A)) + abs(imag(A)) \|_\infty$$

If x is a vector and p is a positive scalar then

$$
\begin{array}{lll}
\texttt{t = norm(x)} & \Rightarrow & t = \|x\|_2 \\
\texttt{t = norm(x,p)} & \Rightarrow & t = \|x\|_p
\end{array}
$$

MATLAB cannot compute the p-norm of a matrix unless $p = 1, 2$, or ∞. Here is a MATLAB segment that estimates the probability that

$$\|Ax\|_2 \ge .90 \|A\|_2 \|x\|_2$$

where A (n-by-n) is given and the entries in x are normally distributed:

```
rand('normal')
c = .9 * norm(A);
count = 0;
for trials = 1:100
    x = rand(n,1);
    if ( norm(A*x) >= c*norm(x) )
        count = count + 1;
    end
end
prob = count/100
```

Largest and Smallest Entries

The largest entry in a real vector v can be computed using **max** :

$$t = \text{max}(v) \qquad \Rightarrow \qquad t = \text{max } v_i$$

Here, v is either a row or column vector. The index of the largest entry can also be computed:

$$[t,k] = \text{max}(v) \Rightarrow \qquad t = v_k = \text{max } v_i$$

If v is a complex vector then

$$w = \text{max}(v) \qquad \Leftrightarrow \qquad w = \text{max}(\text{abs}(v))$$

If A is m-by-n and $m > 1$ (to rule out the row vector case) then

$$v = \text{max}(A) \qquad \Rightarrow \qquad v = [\; \text{max}(A(:,1))\; ,..,\; \text{max}(A(:,n))\;]$$

and

$$[v,i] = \text{max}(A)$$

is equivalent to

```
for j = 1:n
    [v(j),i(j)] = max(A(:,j))
end
```

If A and B have the same size then

$$C = \text{max}(A,B) \qquad \Rightarrow \qquad c_{ij} = \text{max}\{\, a_{ij}, b_{ij}\,\}$$

The function **min** is analogous to **max**. If A is a matrix, then

$$t = \text{max}(\text{min}(A))$$

computes the maximum column minimum.

Sums and Products

If x is a row or column vector of dimension n, then

$$t = \text{sum}(x) \quad \Rightarrow \quad t = x_1 + \ldots + x_n$$

$$t = \text{prod}(x) \quad \Rightarrow \quad t = x_1 \times \cdots \times x_n$$

If A is m-by-n and $m > 1$, then **sum** and **prod** return row vectors. In particular

$$v = \text{sum}(A)$$

is equivalent to

```
for j = 1:n
    v(j) = sum(A(:,j))
end
```

while

$$v = \text{prod}(A)$$

is synonymous with

```
for j = 1:n
    v(j) = prod(A(:,j))
end
```

Thus, if A is a real matrix then

$$t = \text{sqrt}(\text{sum}(\text{sum}(A.*A)))$$

computes the Frobenius norm.

"sort" and "find"

The entries in a real vector can be sorted using **sort**. In particular, if x is a vector then

$$v = \texttt{sort(x)} \quad \Rightarrow \quad v = Px$$

where P is a permutation such that $v_1 \le v_2 \le \cdots \le v_n$. The statement

$$\texttt{[v,i] = sort(x)}$$

returns the sorted vector in v and an integer vector representation of the permutation P. In particular,

$$\texttt{[v,i] = sort(x)} \quad \Rightarrow \quad v(j) = x(i(j)) , \ j = 1{:}n.$$

If x is a column (row) vector then v and i are column (row) vectors. If A is m-by-n and $m > 1$ then

$$\texttt{B = sort(A)} \quad \Rightarrow \quad B = [\, sort(A(:,1)) \,,..., \, sort(A(:,n)) \,]$$

while

$$\texttt{[B,i] = sort(A)} \quad \Rightarrow \quad [\, B(:,j) \,,\, i(j)\,] = sort(\, A(:,j)\,) \ , \ j = 1{:}n$$

If **sort** is applied to a complex argument then it sorts on the absolute value of the argument.

The **find** function can be used to locate nonzero entries. If v is a vector then

$$\texttt{f = find(v)}$$

assigns to f the vector of indices where v is nonzero. Thus,

```
for i = find(v)
    v(i) = v(i) + 1;
end
```

adds one to the nonzero entries in v.

Rounding, Remainders, and Signs

There are several conversion-to-integer routines:

round(x)	$\Rightarrow$	round x to nearest integer
fix(x)	$\Rightarrow$	round x to zero
floor(x)	$\Rightarrow$	round x to $-\infty$
ceil(x)	$\Rightarrow$	round x to $+\infty$

Signs and remainders are also part of MATLAB:

$$t = \text{sign(x)} \quad \Rightarrow \quad t = \begin{cases} 1 & \text{if } x > 0 \\ 0 & \text{if } x = 0 \\ -1 & \text{if } x < 0 \end{cases}$$

$$t = \text{rem(x,y)} \quad \Rightarrow \quad t = x - y \cdot \mathit{fix}(x/y)$$

All of these operations are performed pointwise for nonscalar arguments. Thus,

```
rand('uniform')
A = floor(-ones(n) + 3*rand(n))
```

establishes A as an n-by-n matrix whose entries are randomly selected from the set $\{-1,0,+1\}$.

For complex arguments, the functions **round, fix, floor, ceil, rem,** and **sign** are applied to the absolute value of the argument.

Extracting Triangular and Diagonal Parts

The functions **tril** and **triu** can be used to extract the lower and upper triangular parts of a matrix. If A is a given m-by-n matrix then

B = tril(A) $\Rightarrow$ $b_{ij} = a_{ij}$ if $i \geq j$, zero otherwise
B = triu(A) $\Rightarrow$ $b_{ij} = a_{ij}$ if $i \leq j$, zero otherwise

Other band structures can be achieved using two-argument versions of **triu,** and **tril.** A definition is required. We say that element a_{ij} is on the kth diagonal of A if $j - i = k$. Note that diagonals below the main diagonal are specified by a negative k.

With this definition we have

B = tril(A,k) $\Rightarrow$ $b_{ij} = a_{ij}$ if $i + k \geq j$,   zero otherwise
B = triu(A,k) $\Rightarrow$ $b_{ij} = a_{ij}$ if $i + k \leq j$,   zero otherwise

Thus,

H = triu(A,-1) $\Rightarrow$ H = upper Hessenberg part of A
H = tril(A,1) $\Rightarrow$ H = lower Hessenberg part of A

Likewise,

$$T = \texttt{tril(triu(A,-1),1)}$$

sets T to be the tridiagonal part of A.

The **diag** function can be used to extract diagonals and to set up diagonal matrices. If A is m-by-n then

v = diag(A) $\Rightarrow$ $v = [\, a_{11}, a_{22}, ..., a_{pp} \,]$, $p = min\ \{m,n\}$

while

v = diag(A,k) $\Rightarrow$ $v =$ kth diagonal of A

If x is an n-vector (row or column) then

D = diag(x) $\Rightarrow$ $D = diag(x_1, ..., x_n)$ (n-by-n)

More generally,

$$D = \texttt{diag(x,k)}$$

sets D to be a square matrix of order $n + abs(k)$ with the kth diagonal x and all other diagonals zero.

As an exercise in this notation, if r and c are given n-vectors, then the one-liner

```
        T = toeplitz(r,c)
```

is equivalent to

```
        T = zeros(n)
        for k = 0:n-1
            j = n-k
            if k > 0
                T = T + diag(r(k)*ones(j,1),k)
            end
            T = T + diag(c(k)*ones(j,1),-k)
        end
```

"rank"

If A is m-by-n, then

```
        r = rank(A)
```

computes the numerical rank of A. This computation requires an estimation of A's rank. To this end, MATLAB regards A as a rank r matrix if its singular values satisfy

$$\sigma_1 \geq \ldots \geq \sigma_r > eps\, \sigma_1 \geq \sigma_{r+1} \geq \cdots \geq \sigma_q \geq 0 \qquad q = min\{m,n\}$$

(Recall that **eps** is the unit roundoff.) It is also possible to base the rank decision on an arbitrary tolerance. If *tol* is a nonnegative scalar then

```
        r = rank(A, tol)
```

defines r by

$$\sigma_1 \geq \cdots \geq \sigma_r > tol\, \sigma_1 \geq \sigma_{r+1} \geq \cdots \geq \sigma_q \geq 0$$

Condition

The 2-norm condition σ_1 / σ_{min} of a matrix A can be computed as follows:

```
kappa = cond(A)
```

The singular value decomposition is used to determine the condition. If A has any zero singular values then a call to **cond** prompts the message:

```
condition is infinite
```

A relatively cheap *estimate* of the reciprocal of the 1-norm condition is returned by the function **rcond**:

```
rkappa = rcond(A)
```

Determinant

The determinant of a square matrix A can be computed as follows:

```
d = det(A)
```

Elementary Functions

Elementary functions supported by MATLAB include

trigonometric:	sin(.)	cos(.)	tan(.)
inverse trigonometric:	asin(.)	acos(.)	atan(.)
exponential:	exp(.)	log(.)	log10(.)
hyperbolic:	sinh(.)	cosh(.)	tanh(.)

These functions are performed "pointwise" in the event of a nonscalar argument. Thus,

$$F = \cos(A) \qquad \Rightarrow \qquad f_{ij} = \cos(a_{ij})$$

Functions of Matrices

MATLAB has provisions for computing several important matrix functions. If A is a square matrix then

$$F = \texttt{expm(A)} \quad \Rightarrow \quad F = e^A$$
$$F = \texttt{logm(A)} \quad \Rightarrow \quad A = e^F$$
$$F = \texttt{sqrtm(A)} \quad \Rightarrow \quad F^2 = A$$

More generally, if

$$F = \texttt{funm(A,'\{ \textit{function name} \}')}$$

then F is computed to be the corresponding function of A. The function named must be one of the elementary functions listed above. Thus,

$$F = \texttt{funm(A,'exp')} \quad \Rightarrow \quad F = exp(A)$$

"roots" and "poly"

If A is an n-by-n matrix then

$$\texttt{c = poly(A)} \Rightarrow det(\lambda I - A) = c_{n+1} + c_n\lambda + \cdots + c_2\lambda^{n-1} + c_1\lambda^n$$

The function **poly** can also accept vector arguments. If r is an n-vector then

$$\texttt{c = poly(r)}$$
$$\Leftrightarrow$$
$$c_{n+1} + c_n\lambda + \cdots + c_2\lambda^{n-1} + c_1\lambda^n = (\lambda - r_1) \cdots (\lambda - r_n)$$

Thus,

$$\texttt{c = poly(-ones(1,n))}$$

returns binomial coefficients.

The function **roots** is the inverse of **poly**. If c is an $(n+1)$-vector then

$$r = \text{roots(c)}$$

$$\Leftrightarrow$$

$$(\lambda - r_1) \cdots (\lambda - r_n) = c_{n+1} + c_n\lambda + \cdots + c_2\lambda^{n-1} + c_1\lambda^n$$

Thus,

$$r = \text{roots([a,b,c])}$$

solves the quadratic equation $ax^2 + bx + c = 0$.

Fourier Transforms

If x is an n-vector and n is a power of two then

$$y = \text{fft(x)} \qquad \Rightarrow \qquad y = \text{Fourier transform of } x$$

$$y = \text{ifft(x)} \qquad \Rightarrow \qquad y = \text{inverse Fourier transform of } x$$

Before these functions are applied, the vector x is padded with zeros (if necessary) so that its dimension is a power of two.

The functions **dft** and **idft** correspond to **fft** and **ifft**. However, they do not pad x with zeros if n is not a power of two and they do not invoke the fast Fourier transform.

Two-dimensional Fourier transforms are also possible:

$$B = \text{fft2(A)} \quad \Rightarrow \quad B = \text{2D Fourier transform of } A$$

$$B = \text{ifft2(A)} \quad \Rightarrow \quad B = \text{2D inverse Fourier transform of } A$$

In these operations A is padded with zeros in order to make each of its dimensions a power of two.

Problems for Section 4.4

1. Write a MATLAB segment that checks if the Frobenius norm of the strictly lower triangular portion of a square matrix A is less than $.01 \cdot \|A\|_F$ and if so, zeros the lower triangular portion of A.

2. Write a MATLAB one-liner that overwrites the complex matrix $A = B + iC$ with $C^T - iB^T$. Here, B and C are real.

3. For $c = 0, .1, \cdots, 1.0$ and $n = 10$, compute the condition of the matrix $P_n - ce_ne_n^T$, where P_n is the n-by-n Pascal matrix and e_n is the last column of I_n.

4. Given k-by-k matrices A and B and positive integer p, write a MATLAB one-liner that generates the block tridiagonal matrix:

$$M \quad = \quad \begin{bmatrix} A & B & 0 & 0 & 0 \\ B & A & B & 0 & 0 \\ 0 & B & A & B & 0 \\ 0 & 0 & B & A & B \\ 0 & 0 & 0 & B & A \end{bmatrix} \qquad (p = 5)$$

5. Write a MATLAB one-liner that computes a matrix F so $F^4 = A$, where A is a given n-by-n matrix.

6. Let x be a given n-vector whose entries satisfy $0 \le x_i \le 1$. Write a MATLAB one-liner that sets v to be the vector of indices for which $.25 \le x_i \le .75$.

7. Write a MATLAB segment that overwrites A (m-by-n) with AP, where P is a permutation chosen so that if

$$AP = [a_1, \dots, a_n]$$

is a column partitioning then

$$\|a_1\|_2 \ge \|a_2\|_2 \ge \cdots \ge \|a_n\|_2.$$

8. Let A be an n-by-n symmetric matrix. Write a MATLAB segment that overwrites A with P^TAP, where P is a permutation such that the diagonal entries of P^TAP are monotone decreasing.

9. Suppose A and B are matrices of the same dimension. Write a one-liner that zeros all the a_{ij} that correspond to zero b_{ij}.

4.5 FUNCTIONS

It is possible to define functions in MATLAB, a feature that greatly enhances the power of the system. In this section we use examples to illustrate the various rules associated with user-defined functions. We also cover the commands **disp, eval, nargin, nargout, return,** and **global.**

An Example

A simple example best motivates the ideas behind user-defined functions in MATLAB. Here is a function that returns a unit 2-norm vector in the direction of Az, where A (n-by-n) and z (n-by-1) are given.

Example 4.5-1

```
function  y = prodAx(A,x)
%
%  This function returns a unit 2-norm vector in the
%  direction of Ax, where A is n-by-n and x is n-by-1.
%
    y = A*x;
    c = norm(y);
    if c ~= 0
        y = y/c;
    else
        y = [ 1 ; zeros(length(x)-1,1)];
        disp('Matrix-vector product is zero. First column..
                of I returned.')
    end
```

This function could be used in a MATLAB implementation of the power method. In particular, if A (n-by-n) and $0 \neq v$ (n-by-1) are given then

```
for k = 1:10
    v = prodAx(A,v);
end
```

232

performs 10 steps of the power method, overwriting v with a unit vector in the direction of $A^{10}v$.

The example highlights a number of key points associated with the use of functions.

- The first line in the function definition is of the form

  function {*variable*} = {*function name*} ({*arguments*})

  whenever the function returns a single value. Functions that return values in more than one variable are discussed later.

- Comments begin with the "%" sign. All functions should begin with "how-to-use" comments, as they are displayed when the **help** command is invoked, e.g.,

  help prodAx

  Our indentation of "noncomments" is optional and is done for the sake of readability. Formally, the "%" symbol "says" to ignore the rest of the line. Thus, we could write

  y = y/c ;   **% scale y**

- Output messages can be displayed with commands of the form

  disp (' {*message*} ')

- The variables used by a function that do not appear in the **function** statement are local unless they have been declared global. A discussion of global variables appears later in this section.

- A function can be referred to in an expression so long as it makes dimensional sense, e.g.,

```
w  =  sort (prodAx (A, v))
w  =  prodAx (A, prodAx (A, v))
B  =  prodAx (A, v1) *prodAx (A, v2) '
```

Functions Can Call Other Functions

It is legal for a function to refer to another function as is the case in the following example:

Example 4.5-2 (Normalized Krylov Matrix)

```
function K = kry(A,v,p)
%
%   Computes K = [ v , Av , A^2 v ,..., A^(p-1) v ] D,
%   where A is n-by-n, v is n-by-1, and D is diagonal
%   chosen so that the columns of K have unit 2-norm.
%
    [ m,n ] = size(A);
    nv = length(v)
    if  m  ~= n | n ~= nv
        disp('Dimensions do not agree.')
    else
        K = v/norm(v);
        for j = 2:p-1
            K = [ K  prodAx(A,K(:,j-1)) ];
        end
    end
    return
```

This example illustrates the **return** statement. It essentially has the same role to play in MATLAB as it does in Fortran. A MATLAB function can have more than one return or it need not have any.

Multiple Input and Output Arguments

A function can return more than one matrix, vector, or scalar value. We have already seen this with some of the built-in functions such as **sort, sum, prod**, etc. Here is a user-defined function with the same property:

Example 4.5-3

```
        function [P N] = pn(A)
%
% Computes nonnegative matrices P and N, where
% A(i,j) > EPS  implies P(i,j) = A(i,j) , N(i,j) = 0
% A(i,j) < EPS  implies P(i,j) = 0, N(i,j) = -A(i,j).
% | A(i,j) | <= EPS implies P(i,j) = N(i,j) = 0
%
    E = EPS*ones(A);
    P =  A.*( A > E );
    N = -A.*(-A > E ) ;
    end
```

Now suppose we want to increase the flexibility of the function **pn** in two ways. First, we wish to be able to invoke a user-defined tolerance *tol* that can be used instead of the machine precision. Second, we wish to be able to return *P* alone if that is all that is required. In other words, we would like **pn** to handle the four possibilities:

Statement	Tolerance	Output
P = pn(A)	eps	Just P
[P,N] = pn(A)	eps	P and N
P = pn(A,tol)	tol	Just P
[P,N] = pn(A,tol)	tol	P and N

These possibilities can be achieved by using **nargin** and **nargout**, special variables that indicate the number of input and output arguments associated with a given function call. The called function can examine **nargin** and **nargout** and proceed accordingly. For example, the built-in function **norm** uses **nargin** to sense whether a 2-norm is to be computed (alfa = norm(x)) or a *p*-norm (alfa = norm(x,p)). Likewise, **sort** uses **nargout** to determine whether to return the sorted vector (y = sort(x)) or the sorted vector plus index array ([y idx] = sort(x)). Here is how **nargin** and **nargout** can be used to set up the function **pn** as outlined in the above table.

Example 4.5-4

```
    function [P N] = pn(A,tol)
%
% Computes nonnegative matrices P and N, where
% A(i,j) > EPS  implies P(i,j) = A(i,j) , N(i,j) = 0
% A(i,j) < EPS  implies P(i,j) = 0, N(i,j) = -A(i,j).
% | A(i,j) | <= EPS implies P(i,j) = N(i,j) = 0
%  If one input argument, tol = EPS.
%  If one output argument, P only is returned.
%
    if nargin == 1
        tol = EPS;
    end
    E = tol*ones(A);
    P = A.*( A > E );
    if nargout == 2
        N = -A.*(-A > E ) ;
    end
```

Passing Function Names

It is possible to pass the name of a function to another function, but it requires some text processing with the **eval** function. Here is an example:

Example 4.5-5

```
      function F = afun(A,f)
%
%  If A is an m-by-n matrix and f is the name of either
%  a built-in or user-defined function that operates
%  on scalars, then this function returns an m-by-n
%  matrix F with F(i,j) = f(A(i,j)). The name of f must
%  be in quotes, e.g., F = afun(A,'cos')  sets F(i,j)
%  to  cos(A(i,j)) .
%
    [m,n] = size(A);


    for j = 1:m
```

```
      for i = 1:n
          F(i,j) = eval([ f ,  '(A(i,j))' ] );
      end
   end
```

To understand this function we need to understand the statement

```
   F(i,j)  =  eval([ f ,  '(A(i,j))' ] )
```

which clearly must assign $f(A(i,j))$ to $F(i,j)$.

A string in MATLAB is a row vector whose entries are characters. Strings are specified by enclosing them in quotes. Thus

```
   s1 = 'cos'
```

assigns the three-character string 'cos' to the string $s1$. Strings can be concatenated. Thus,

```
   s1 = 'cos'
   s2 = '(A(i,j))'
   s = [ s1 s2]
```

is equivalent to

```
   s = 'cos(A(i,j))'
```

The built-in function **eval** takes a string and executes it, assuming that the string is a meaningful MATLAB statement. Thus,

```
   eval( 'y = A*x')
```

is equivalent to

```
   y = A*x
```

Returning to our Example 4.5-5, the argument f is a string and so the statement

```
eval( [ f '(A(i,j))' ] )
```

means: concatenate the contents of `f` with `'(A(i,j))'` and execute. If `f` has the value `'cos'` we see that

```
F(i,j) = eval([ f , '(A(i,j))' ] )
```

is equivalent to

```
F(i,j) = cos(A(i,j))
```

This is precisely what is required.

Global Variables

The MATLAB version of the Fortran common construct is **global**. The values in all **global** variables can be accessed whenever inside a function. Suppose x and y are initialized vectors of equal dimension and we execute

```
global y
a = f(x)
```

where

```
function a = f(z)
a = y'*z
```

This is equivalent to

```
a = y'*x
```

Like **common** in Fortran, **global** should be used sparingly.

Recursion

A MATLAB function can refer to itself. We illustrate this with a somewhat contrived example that is concerned with L_1-norm computation.

Observe that if x is an n-vector then

$$\| x \|_1 = \| x(1{:}n{-}1) \|_1 + | x(n) |$$

This recursive definition forms the basis of the following function:

Example 4.5-6

```
function normL1 = f(x)
%
%  Returns the 1-norm of an n-vector x
%
    n = length(x);
    if n == 1
        normL1 = abs(x);
    else
        normL1 = f( x(1:n-1) ) + abs(x(n));
    end
```

Note that **f** must be called n times to compute the 1-norm of an n-vector.

A more interesting example of recursion is the Strassen matrix multiply procedure. Consider 2-by-2 block matrix multiplication:

$$\begin{bmatrix} C_{11} & C_{12} \\ C_{21} & C_{22} \end{bmatrix} = \begin{bmatrix} A_{11} & A_{12} \\ A_{21} & A_{22} \end{bmatrix} \begin{bmatrix} B_{11} & B_{12} \\ B_{21} & B_{22} \end{bmatrix}$$

Assume that all the blocks are square and of the same dimension. Strassen showed how to compute C with *seven* multiplies and *eighteen* adds :

$$\begin{aligned} P_1 &= (A_{11} + A_{22})(B_{11} + B_{22}) \\ P_2 &= (A_{21} + A_{22}) B_{11} \\ P_3 &= A_{11} (B_{12} - B_{22}) \\ P_4 &= A_{22} (B_{21} - B_{11}) \\ P_5 &= (A_{11} + A_{12}) B_{22} \\ P_6 &= (A_{21} - A_{11})(B_{11} + B_{12}) \\ P_7 &= (A_{12} - A_{22})(B_{21} + B_2) \end{aligned}$$

$$
\begin{aligned}
C_{11} &= P_1 + P_4 - P_5 + P_7 \\
C_{12} &= P_3 + P_5 \\
C_{21} &= P_2 + P_4 \\
C_{22} &= P_1 + P_3 - P_2 + P_6
\end{aligned}
$$

These equations are easily confirmed by substitution. The Strassen algorithm can also be applied to each of the half-sized block multiplications associated with the P_i. Thus, if the original A and B are n-by-n, and $n = 2^q$ we can apply Strassen q times. At the bottom "level," the blocks are 1-by-1. Of course, there is no need to recur until the bitter end. When block size gets sufficiently small, ($n \leq nmin$), it is smart to use conventional matrix multiplication when finding the P_i.

Example 4.5-7

```
function C = strass(A,B,nmin)
%
%  Strassen matrix multiplication C = AB, A,B square.
%  If n <= nmin, multiply done conventionally.
%
   [ n n ] = size(A) ;
   if  n <= nmin
       C = A*B;     % n small , get C conventionally
   else
       m = n/2; u = 1:m ; v = m+1:n;
       P1 = strass(A(u,u)+A(v,v),B(u,u)+B(v,v) , nmin );
       P2 = strass( A(v,u) + A(v,v) , B(u,u) , nmin );
       P3 = strass( A(u,u) ,  B(u,v) - B(v,v) , nmin );
       P4 = strass( A(v,v) , B(v,u) - B(u,u) , nmin );
       P5 = strass( A(u,u) + A(u,v) , B(v,v) , nmin ) ;
       P6 = strass( A(v,u)-A(u,u) ,B(u,u)+B(u,v) , nmin );
       P7 = strass( A(u,v) - A(v,v),B(v,u)+B(v,v),nmin );
       C = [ P1+P4-P5+P7   P3+P5 ; P2+P4    P1+P3-P2+P6 ];
   end
```

Problems for Section 4.5

1. Write a MATLAB function

$$A = prod3(B, C, D)$$

that computes the product $A = BCD$, where B, C, and D are matrices. The order of multiplication should be determined to minimize the number of required flops. An error message should be printed if either (BC) or (CD) is not defined because of dimension conflicts.

2. Write a MATLAB function

$$C = longprod(A, B, s)$$

where A and B are square matrices of the same size and s is a string that prescribes the product of A's and B's that is to be performed. For example

$$C = longprod(X, Z, 'XXZXZZ')$$

would be equivalent to

$$C = X*X*Z*X*Z*Z$$

Assume that the names of the two input matrices are one character long.

3. Write a MATLAB segment that generates twenty 2-by-2 random matrices and stores them in A00, A01,..., A19. Hint: Set $s = '0123456789'$ then generate matrices using **eval** and concatenation.

4. Modify the function **strass** so that it can handle arbitrary n. Do this by enlarging the "current" A and B if they are odd-dimensioned.

5. Write a MATLAB function $[L , m] = sylv(X,A,B)$ that performs as follows:

$$[L, m] = sylv(X, A, B) \implies L = AX + XB, \quad m = \| L \|_2$$
$$L = sylv(X, A, B) \implies L = AX + XB$$

```
[L,m]  =  sylv(X,A)       ⇒   L = AX + XA^T ,   m = ||L||_2
L  =  sylv(X,A)           ⇒   L = AX + XA^T
```

$$[L,m] = \text{sylv}(X,A) \implies L = AX + XA^T, \quad m = ||L||_2$$
$$L = \text{sylv}(X,A) \implies L = AX + XA^T$$

Check to make sure that the dimensions of A, B, and X make sense.

4.6 FACTORIZATION

One of the most powerful aspects of MATLAB is the ease with which various matrix factorizations can be computed. In this section we survey the high-level commands that make these computations so easy.

Solving Linear Systems

If A (n-by-n) and B (n-by-p) are given, then

```
X  =  A\B
```

computes the solution to the multiple right-hand side linear system $AX = B$. A message appears if A is too ill-conditioned. The backslash operator invokes Gaussian elimination with partial pivoting.

MATLAB does *not* exploit any sparsity structure that may be present in the matrix A when it computes $A\backslash B$. Thus, eye (B) \B requires just as much work as $A\backslash B$ for general A.

MATLAB is also able to compute inverses:

```
Ainv  =  inv(A)
```

Thus, a less preferable way to solve $AX = B$ is

```
X  =  inv(A)*B
```

The LU Factorization

The factorization $A = LU$, where U is upper triangular and L is a row-permuted unit lower triangular matrix, can be computed as follows:

```
[ L,U ]  =  lu(A)
```

Thus,

```
[L,U]  =  lu(A);
y  =  L\b;
x  =  U\y;
```

is a much less efficient method for solving $Ax = b$ than the equivalent

```
x  =  A\b
```

Only square matrices can be factored by **lu**.

Note that **lu** does not compute the "true" LU factorization $PA = LU$, where P is a permutation, L is lower triangular, and U is upper triangular.

The Cholesky Factorization

If A is an m-by-n Hermitian, positive definite matrix, then

```
R  =  chol(A)
```

computes an upper triangular R such that $A = R^H R$. Only the upper triangular portion of A is accessed.

There is no built-in function for computing the factorization $A = LDL^H$, where L is unit lower triangular and D is diagonal. However, the segment

```
L  =  chol(A)';
d  =  diag(L);
L  =  L'
for  k  =  1:n
    L(:,k)  =L(:,k)/d(k);
end
D  =  diag( d.*d);
```

does the job.

QR Factorization

If A is an m-by-n matrix then

```
[ Q,R ] = qr(A)
```

computes unitary Q (m-by-m) and upper triangular R (m-by-n) such that $A = QR$. If b (m-by-1) is given and A has full column rank then

```
[ Q,R ] = qr(A);
[m,n] = size(A);
b = Q'* b ;
x = R(1:n,1:n) \ b(1:n) ;
```

computes the minimizer of $||Ax - b||_2$.

Householder and Givens Transformations

If x (n-by-1) is initialized then the following instruction computes a Householder matrix P such that Px is zero below the first component:

```
[ P , R ] = qr(x)
```

This works because **qr** invokes the Householder upper triangularization algorithm.

If x is a 2-by-1 vector then

```
[ Q , R ] = qr(x);
c = Q(1,1);
s = -Q(2,1) ;
```

computes a cosine-sine pair (c,s) such that

$$
\begin{bmatrix} c & s \\ -s & c \end{bmatrix} \begin{bmatrix} x_1 \\ x_2 \end{bmatrix} = \begin{bmatrix} r \\ 0 \end{bmatrix}
$$

QR Factorization with Column Pivoting

If A is an m-by-n matrix then

```
[ Q,R,P ] = qr(A)
```

computes unitary Q (m-by-m), permutation P (n-by-n), and upper triangular R (m-by-n) so that

$$Q^T A P = R = \begin{bmatrix} R_{11} & R_{12} \\ O & 0 \end{bmatrix}$$

Here, R_{11} is r-by-r and R_{12} is r-by-$(n-r)$, where $r = rank(A)$. Moreover, for $i = 1:min\{m,n\}$ we have

$$|r_{ii}|^2 \geq |r_{ij}|^2 + ... + |r_{jj}|^2 \qquad j = i,..., n$$

If b (m-by-1) and A (m-by-n) are given with $n < m = rank(A)$, then

```
[ Q,R,P ] = qr(A) ;
[ m,n ] = size(A);
y = R(1:m , 1:m) \ Q'*b ;
x = P * [y ; zeros(n-m,1) ] ;
```

produces a solution to the underdetermined system $Ax = b$.

Range and Null Space

If A (m-by-n) is given and has full rank, then an orthonormal basis for the range space can be computed as follows:

```
[ m,n ] = size(A);
[ Q,R ] = qr(A);
Q = Q(:,1:n);
```

Alternatively, Q can be computed using the built-in function **orth**:

```
Q = orth(A);
```

If A is rank deficient the number of columns in Q will equal $rank(A)$.

An orthonormal basis for the null space of a matrix can be obtained using **null**. If

$$Q = \text{null}(A)$$

then $Q^T Q = I_k$ and $AQ = 0$, where $k = dim(null(A))$. If A has full column rank then **null** returns the empty matrix.

The Singular Value Decomposition

If A (m-by-n) is given then

$$[U, S, V] = \text{svd}(A)$$

computes unitary U (m-by-m), unitary V (n-by-n), and diagonal S (m-by-n) so

$$U^H A V = S$$

This is the singular value decomposition and the diagonal entries of S are arranged so that

$$s_{11} \geq s_{22} \geq \cdots \geq s_{qq} \geq 0 \qquad q = min\{m,n\}$$

If $m \geq n$ then the "compact" SVD in which U is m-by-n and S is n-by-n can be determined using a two-argument version of **svd**:

$$[U, S, V] = \text{svd}(A, 0);$$

This is equivalent to

```
[ U, S, V ] = svd(A);
[ m, n ] = size(A);
if m >= n
    U = U(:, 1:n);
    S = diag(diag(S));
end
```

If all that is required is the vector of singular values then

```
sv = svd(A);
```

is a shortcut for

```
[ U,S,V ] = svd(A);
sv = diag(S);
```

The function **svd** returns the singular values in largest-to-smallest order. Thus, if $p \leq rank(A)$ then

```
[ U,S,V ] = svd(A);
AP = U(:,1:p)*S(1:p,1:p)*V(:,1:p)' ;
```

computes the closest rank p matrix to A, while

```
[ U,S,V ] = svd(A,0) ;
r = rank(A) ;
b = V(:,1:r)*( S(1:r,1:r) \ (U(:,1:r)'*b));
```

overwrites b with the minimum 2-norm minimizer of $||Ax - b||_2$.

Pseudoinverse and Least Squares

If A (m-by-n) is initialized then the pseudoinverse X (n-by-m) can be found using **pinv**:

```
X = pinv(A)
```

Thus, the minimum 2-norm minimizer of $||Ax - b||_2$ can be found as follows:

```
x = pinv(A)*b ;
```

The backslash operator can also be used to solve least squares problems. However,

```
x = A\b ;
```

computes the basic solution via QR with column pivoting. Multiple right-hand side least squares problems can be solved with either of the commands

```
X = A\B;
```
or
```
X = pinv(A)*B;
```

Here, B is an m-by-p matrix.

The Hessenberg Decomposition

If A (n-by-n) is given then

```
[ Q,H ] = hess(A);
```

computes unitary Q (n-by-n) and upper Hessenberg H such that $Q^H A Q = H$. The instruction

```
H = hess(A)
```

assigns the Hessenberg form A to H.

As an example of **hess**, here is a segment that computes a Hessenberg decomposition of A subject to the constraint that the first column of Q is a unit 2-norm vector in the direction of a given vector b :

```
[ Q1 , R ] = qr( b );
[ Q, H ] = hess( Q1'*A*Q1);
Q = Q1*Q;
```

In practice, one would want to exploit the fact that the matrix $Q1$ is a Householder matrix.

The Schur Decomposition

The statement

```
[ Q,T ] = schur(A);
```

computes the Schur form of the n-by-n matrix A. There are two cases to distinguish. If A is real then Q is orthogonal and $T = Q^TAQ$ is upper quasi-triangular. This is the real Schur form of A. If the imaginary part of A is nonzero, then Q is unitary and $T = Q^HAQ$ is upper triangular.

The Schur decomposition of a real symmetric matrix A has the form

$$Q^TAQ = diag(d_1,...,d_n)$$

where the eigenvalues can be ordered as follows:

$$d_1 \geq d_2 \geq \cdots \geq d_n$$

This can be obtained as follows:

```
[Q  , D ] = schur(A);
[ D1 , indx ] = sort(diag(D));
indx = indx( n:-1:1);
Q = Q( : , indx );
D = D(indx,indx);
```

The function **rsf2csf** can be used to compute the complex Schur form from the real Schur form. (Recall that a real matrix can have complex eigenvalues in which case a complex unitary Q must be used to upper triangularize A.) In particular, if A is square and

```
[ Q , T ] = schur(A);
[ Q , T ] = rsf2csf(Q,T);
```

then Q unitary and $Q^HAQ = T$ is upper triangular.

The Eigenvector Decomposition

If A is given then

```
[ X,D ] = eig(A);
```

produces a nonsingular X and diagonal D such that $X^{-1}AX = D$ is diagonal. If A has a defective eigensystem then X will be ill conditioned or may even have some repeated columns.

Recall that the condition of a simple eigenvalue is the secant of the angle between the left and right eigenspaces. The vector of reciprocal eigenvalue condition numbers $(s_1,...,s_n)$ can be obtained as follows:

```
Y = inv(X);
for i = 1:n
    xi = X(:,i:i);
    yi = conj(Y(i:i,:));
    s(i) = abs (xi*yi)/(norm(xi)*norm(yi));
end
```

The vector of eigenvalues of a square matrix A can be obtained as follows:

```
ev = eig(A);
```

Thus,

```
rho = max(abs(eig(A)));
```

computes the spectral radius of the matrix A, i.e., the magnitude of the largest eigenvalue in absolute value. Likewise,

```
alfa = max(real(eig(A)))
```

computes the spectral abscissa, the maximum real part of any eigenvalue.

Generalized Eigenvalues

A two-argument version of **eig** can be used to solve the generalized eigenvalue problem

$$Ax = \lambda Bx$$

where A and B are square and of the same dimension. In particular,

```
[X,D] = eig(A,B)
```

computes a matrix X (possibly ill conditioned) and a diagonal matrix D such that

$$AX = BXD$$

Note that if $X = [x_1, ..., x_n]$ and $D = diag(d_1, ..., d_n)$ then $Ax_i = d_i Bx_i$, where $i = 1{:}n$. Entering

```
ev = eig(A,B)
```

sets ev to be the vector of generalized eigenvalues.

Because it is possible to have infinite generalized eigenvalues and related awkward behavior, it is frequently preferable to compute the QZ decomposition of the pair (A,B) rather than its eigenvalue decomposition. In particular,

```
[ R,S,Q,Z,X ] = qz(A,B)
```

computes unitary Q and Z such that $Q^H A Z = R$ and $Q^H B Z = S$ are upper triangular. The matrix $X = [x_1, ..., x_n]$ is the matrix of generalized eigenvectors and satisfies

$$r_{ii} Ax_i = s_{ii} Bx_i$$

for $i = 1{:}n$. The function **eig** uses **qz** in the sense that

```
[ R,S,Q,Z,X ] = qz(A,B);
D = diag(diag(R)./diag(S));
```

renders the same diagonal D as

$$D = \text{eig}(A,B).$$

Problems for Section 4.6

1. Given A (p-by-p), B (p-by-q), C (q-by-q), g (p-by-1), and h (q-by-1), write a MATLAB segment that solves the linear system

$$\begin{bmatrix} A & B \\ O & C \end{bmatrix} \begin{bmatrix} y \\ z \end{bmatrix} = \begin{bmatrix} g \\ h \end{bmatrix}$$

2. Suppose

$$C = \begin{bmatrix} A & v \\ v^T & \alpha \end{bmatrix}, \quad v \in R^{n-1}$$

is positive definite and that R is an upper triangular matrix such that $A = R^T R$. Write an efficient MATLAB segment that overwrites R with C's Cholesky factor.

3. Let A and B be symmetric with B positive definite. Write a MATLAB segment that computes the n generalized eigenvalues of the problem

$$Ax = \lambda Bx$$

by computing the eigenvalues of $C = R^{-T} A R^{-1}$, where $B = R^T R$ is the Cholesky factorization of B.

4. Suppose A (m-by-n) is given with with $m \geq n$. Write a MATLAB segment that computes a row-permuted unit lower triangular matrix L (m-by-n) and an upper triangular U (n-by-n) such that $A = LU$.

5. Write a MATLAB segment that overwrites A (n-by-n) with

$$Q^H A Q = \begin{bmatrix} \lambda & w^H \\ 0 & B \end{bmatrix}$$

where Q is unitary and λ has the largest absolute value amongst A's eigenvalues.

6. Write a MATLAB function

$$[Q, R] = qrmod(A, k)$$

that computes one of the following factorizations:

$$
\begin{array}{lll}
k = 1 & \Rightarrow \ A = QR, & Q \text{ unitary, } R \text{ upper triangular} \\
k = 2 & \Rightarrow \ A = QR, & Q \text{ unitary, } R \text{ lower triangular} \\
k = 3 & \Rightarrow \ A = RQ, & Q \text{ unitary, } R \text{ upper triangular} \\
k = 4 & \Rightarrow \ A = RQ, & Q \text{ unitary, } R \text{ lower triangular}
\end{array}
$$

Make use of **qr** and the fact that if $E = I(:, n:-1:1)$ is the n-by-n *exchange matrix* then

$$
\begin{array}{ll}
Q \text{ unitary} & \Rightarrow \ EQE \text{ unitary} \\
T \text{ upper triangular} & \Rightarrow \ ETE \text{ lower triangular}
\end{array}
$$

Assume that A is square.

7. Note that if

$$A = \begin{bmatrix} B & v \\ v^H & \alpha \end{bmatrix}$$

is symmetric positive definite then

$$chol(A) = \begin{bmatrix} chol(B) & w \\ 0 & \beta \end{bmatrix}$$

where $chol(B)^T w = v$ and $\beta = sqrt(\alpha - w^T w)$. Use this to develop a recursive MATLAB function

```
R = cholrecur(A)
```

that returns A's Cholesky factor.

4.7   MISCELLANEOUS

In this concluding section we cover a selection of topics that assist in the smooth and effective running of MATLAB.

IEEE Arithmetic

The result of a zero divide is a special value called "inf." Thus,

```
x = 1/0
```

prompts the response

```
Warning: Divide by zero.
x = inf
```

This is much more convenient than a fatal overflow exception. Generation of an "inf" usually implies some miscalculation but this need not be the case. Consider the following MATLAB segment that confirms an instance of the Holder inequality

$$|x^H y| \leq ||x||_p ||y||_q$$

for $p = 1:10$:

```
for p = 1:10
    q = p/(p -1);
    alfa = norm(x,p)  *  norm(x,q)   -  abs(x'*y)
end
```

If $p = 1$ and $q = \infty$ then

```
w = norm(x,q)           ⟺        w = norm(x,inf)
```

as required.

Quotients of the form $0/0$ produce a different type of result. In particular, the statement

```
x = 0/0
```

prompts the response

```
Warning: Divide by zero.
x = NaN
```

Here, "NaN" means "not a number."

NaNs and infs can appear in expressions. Thus if $x = inf$ and $y = NaN$ then

`a = x + z`	$\Rightarrow$	$a = inf$
`a = 1/x`	$\Rightarrow$	$a = 0$
`a = y*z`	$\Rightarrow$	$a = NaN$
`a = x / y`	$\Rightarrow$	$a = NaN$
`a = [ x ; y ]`	$\Rightarrow$	$a(1) = inf, a(2) = NaN$

Timings

It is possible to time executions using **clock** and **etime**. Here is a MATLAB segment that records the time required for the matrix vector multiplication $y = Ax$, where A is n-by-n and $n = 5:5:50$:

```
for k = 1:10
    n = 5*k;
    A = rand(n);
    x = rand(n,1);
    t1 = clock;
    y = A*x;
    t2 = clock;
    time(k) = etime(t2,t1);
end
```

The function **clock** returns the current time in a row vector having dimension six. (It specifies the year, month, day, hour, minute, and second.) The "difference" between the two "time vectors" t1 and t2 is computed by **etime**, which returns the elapsed time in seconds. Such computations are fairly accurate at the millisecond level.

Timed Displays

A delay can be built into a MATLAB segment with the **pause** statement. Here is an example that displays the *n*-by-*n* magic square matrix on the screen for approximately *n* seconds for n = 1:15 .

```
for n = 1:15
   A = magic(n)
   pause(n)
end
```

Clearly, pause (n) establishes a pause of *n* seconds. To freeze execution until any key is struck, just use **pause** without argument. For example,

```
for n = 1:15
   A = magic(n)
   disp('Strike any key to see next..
            magic square')
   pause
end
```

permits arbitrarily long scrutiny of the first 15 magic squares.

Getting Input

It is possible to prompt the user for an input value during execution of a MATLAB segment. For example,

```
n = 1
while (n >= 1)
   A = magic(n)
   n = input('Enter dimension')
end
```

displays magic squares of chosen dimension. The process terminates as soon as a nonpositive dimension is specified.

Multiple assignment can also be used in conjunction with input, e.g.,

```
[m,n]  =  input ('Enter  row/column  dimensions.')
```

Converting Numbers to Strings

Recall that a string is a row vector whose *k*th entry is the *k*th character in the string (suitably encoded). The function **num2str** can be used to convert a number to its string format, which can then participate in string operations such as concatenation. Here are some examples of **num2str:**

```
s  =  num2str(2)              ⇒      s   =  '2.0000'
s  =  num2str(-100)           ⇒      s   =  '-1.0000E+2'
s  =  num2str(12345678)       ⇒      s   =  '1.2345E+7'
```

Problems for Section 4.7

1. Write a MATLAB function

```
time  =  bench(n,f)
```

that returns the time it takes to compute f (rand(n)). Here *f* is a string that names any MATLAB function that can operate on a square matrix. Sample:

```
time  =  bench(10,'svd').
```

2. Write a MATLAB function

```
y  =  aqx(A,x,t)
```

that computes the vector $y = A^q x$, where x (*n*-by-1), A (*n*-by-*n*), and $q = 2^t$. Let A = rand(5) and x = rand(5,1). Print T(1:10) with the property that T(t) is the length of time required to execute aqx(A,x,t). Hint: there is more to this problem than just matrix-vector multiplication. Your final program should not be tailored to this particular example. How you compute $A^q x$ depends on n and q, and you are expected to work out a heuristic that "picks" the appropriate method as a function of these two parameters.

References

[1] J. Dongarra, J.R. Bunch, C.B. Moler, and G.W. Stewart (1978), LINPACK *User's Guide,* Society for Industrial and Applied Mathematics, Philadelphia, PA.

[2] J. Dongarra, J. Du Croz, S. Hammarling, and R.J. Hanson (1984), *An extended set of Fortran Basic Linear Algebra Subprograms,* Report TM41, MCS Division, Argonne National Laboratory, Argonne, IL; ACM *Trans. Math. Software,* to appear.

[3] J. Dongarra, J. Du Croz, I.S. Duff, and S. Hammarling (1987), *A proposal for a set of level 3 Basic Linear Algebra Subprograms,* Report TM 212, MCS Division, Argonne National Laboratory, Argonne, IL.

[4] G.H. Golub and C. Van Loan (1983), *Matrix Computations,* The Johns Hopkins University Press, Baltimore, MD.

[5] D. Kahaner, C.B. Moler, and S. Nash (1988), *Numerical Methods and Software,* Prentice-Hall, Englewood Cliffs, NJ.

[6] C.L. Lawson, R.J. Hanson, F.T. Krogh, and O.R. Kincaid (1979), *Basic Linear Algebra Subprograms for FORTRAN Usage,* ACM *Trans. Math. Software,* Vol. 5, pp. 308-323.

[7] C.B. Moler, J. Little, S. Bangert, and S. Kleiman (1987), *ProMatlab User's Guide,* MathWorks, Sherborn, MA.

[8] V. Zwass (1981), *Programming in FORTRAN,* Barnes and Noble, New York, NY.

Index

Fortran 77

BLAS

LINPACK

MATLAB